Understanding Complexity

Understanding Complexity

Edited by

Gillian Ragsdell
University of Paisley
Paisley, Scotland

and

Jennifer Wilby
University of York
York, England
and
Centre for Systems Studies
Hull, England

A Commemorative Volume of the World Congress for the Systems Sciences and ISSS 2000, Toronto, Canada

Springer Science+Business Media, LLC

Library of Congress Cataloging-in-Publication Data

Understanding complexity/edited by Gillian Ragsdell and Jennifer Wilby.
p. cm
Includes bibliographical references and index.
ISBN 978-1-4613-5492-5 ISBN 978-1-4615-1313-1 (eBook)
DOI 10.1007/978-1-4615-1313-1
1. System theory—Congresses. 2. Computational complexity—Congresses. I. Ragsdell, Gillian, 1962- II. Wilby, Jennifer, 1953-

Q295 .U66 2001
003—dc21

2001029580

ISBN 978-1-4613-5492-5

Originally published by Kluwer Academic / Plenum Publishers, New York in 2001
Softcover reprint of the hardcover 1st edition 2001
http://www.wkap.nl

10 9 8 7 6 5 4 3 2 1

A C.I.P. record for this book is available from the Library of Congress

FOREWORD

Peter A. Corning

Palo Alto, CA
November, 2000

This volume represents a distillation of the plenary sessions at a unique millennium-year event - a World Congress of the Systems Sciences in conjunction with the 44th annual meeting of the International Society for the Systems Sciences (ISSS). The overall theme of the conference was "Understanding Complexity in the New Millennium." Held at Ryerson Polytechnic University in Toronto, Canada, from July 16-22, 2000, the conference included some 350 participants from over 30 countries, many of whom were representatives of the 21 organizations and groups that co-hosted this landmark event. Each of these co-host organizations/groups also presented a segment of the program, including a plenary speech.

In addition, the conference featured a number of distinguished "keynote" speeches related to the three daily World Congress themes: (1) The Evolution of Complex Systems, (2) The Dynamics of Complex Systems, and (3) Human Systems in the 21st Century. There were also seven special plenary-level symposia on a range of timely topics, including: "The Art and Science of Forecasting in the Age of Global Warming"; "Capitalism in the New Millennium: The Challenge of Sustainability"; "The Future of the Systems Sciences"; "Global Issues in the New Millennium"; "Resources and the Environment in the New Millennium"; "The Lessons of Y2K"; and "Can There be a Reconciliation Between Science and Religion?"

Included in this special commemorative volume is a cross-section of these presentations. All of the plenary speakers were invited to contribute written summaries of their talks, but some were not able to do so. (Breakout panel papers were published separately in the regular ISSS conference Proceedings.) Nevertheless, the presentations included here represent a permanent record -- along with the Webcast -- of an important and memorable occasion. We are pleased to know that, through this commemorative volume, we will be able to reach an even broader audience. On behalf of the ISSS, which organized the event, I wish to thank all of the participants and especially the many outstanding speakers. They more than fulfilled our high hopes for this meeting.

PREFACE

Gillian Ragsdell
University of Paisley, Paisley, UK

Jennifer Wilby
University of York, York, UK

To be given the opportunity to bring together a selection of highly regarded academic papers is always a privilege. Bringing this particular selection together is a special privilege given the significance of the volume. Firstly, it is a volume that has an important role to play in capturing the essence of a unique millennium-year event – a World Congress of the Systems Sciences in conjunction with the 44th Annual Meeting of the International Society for Systems Sciences (ISSS). Secondly, it is a volume that can stand alone as a respected collection of works from key contributors to the systems discipline in the 20th century. However, as well as being a record of recent systems interests and foci, the volume has an influential role to play in the 21st century for the issues and concerns raised in the papers will, no doubt, guide and shape the future of the systems discipline.

While the invitation to edit this volume represented a privilege, the actual process of editing soon became a challenge! For how does one organise a collection of papers on such diverse topics as the evolution of dynamic systems, human systems, and the art and science of forecasting so that there is coherence and clarity in presentation? How does one retain an author's individual style (compounded by the fact that these papers have emerged from keynote addresses, presidential speeches and plenary sessions), while at the same time, ensure that the volume conveys a sense of 'wholeness'? In the same vein, how does one organise such diverse papers so that a similarly diverse readership is able to experience a meaningful 'story line'? These, and other difficult issues, were ones with which we grappled during the development of this volume. The final arrangement of the papers was chosen so as to align as closely as possible with the main conference themes although observant conference attendees will realise that we have relocated a handful of papers out of their original conference section! It is the final arrangement that is now discussed.

The volume has been organised into eight sections, each providing valuable arguments and discussion points for furthering our understanding of complexity. The opening section of this volume is entitled 'Theories for Complex Systems' and comprises five papers. It is very fitting that the first paper should be authored by one of the founders of General Systems Research. Rapoport's paper, entitled 'The Systemic View of the World: The ∫ Logo of General Systems Research' begins this section. There then follow four other papers that are more specific in their theories. In 'Systems Science Fundamental Principles', Simms gives proofs of fundamental principles for understanding living and non-living systems, with Chaisson putting forward cosmic evolution as an intellectual basis for understanding environmental conditions needed for material assemblages to have become increasingly complex, in 'A Cosmological Imperative: Emergence of Information, Complexity and Life.' Sabelli's 'The Co-Creation Hypothesis' centres on a general hypothesis that draws from empirical studies and mathematical models of creative processes, while Salthe's paper 'What is Infodynamics?' closes this section with a discussion of structural attractors influencing system development.

The four papers in the second section represent some 'Emerging Foci on Systems Research and Practice'. Three of the contributions reflect emerging aspirations from an organisational setting - their titles are relatively self-explanatory. Lee's paper starts this futuristic look at systems work and is entitled 'The History, Goals, Accomplishments and Future plans of the Research Committee on Sociocybernetics of the International Sociological Association' (RC51). In 'Systems Practice at the United Kingdom's Open University, Ison follows on with discussion of the emphasis on applied teaching and research undertaken in the Systems Discipline at the OU. This section then takes a cybernetic slant with Vallée's paper 'The WOSC and some Views about Cybernetics. The history of WOSC is presented before Vallee expresses a preference for the purpose of cybernetics to be the provision of an epistemopraxiological model of how much a conscious being is able to know and modify both itself and the rest of the world. Beer's speech for the 'Presentation of Wiener Gold Medal to Dr. Candace Pert' draws this section to a close. His speech is in keeping with the other three contributions on future aspects of systems work. Beer celebrates Candace's work and predicts that her contribution will help to cross the East-West divide and the chasm that exists between science and philosophy.

It should perhaps not be too surprising that the greatest number of papers fell squarely into the section based on 'Human Systems in the 21st Century', the third section of this volume; after all, systems practice is often concerned with improving situations both for and with people. Ackoff and Pourdehnad's paper with the title 'On Misdirected Systems' bases a clear discussion around issues related to 'doing the right thing wrong' or 'doing the wrong thing right'. Friend implicitly follows on the theme of 'doing the right thing right' in his paper 'Engaging with Transient Complexity in Development Projects'. Drawing from his experience in operations research (OR), action research (AR) and participative development (PD) he moves through the evolution of a decision-centre approach to that of complexity as a platform to review some of the prospects for future synthesis with particular reference to the management of international development projects. The section continues with Zwick's 'Understanding Imperfection', a paper that works through both negative and positive assertions about the uses of systems ideas, with emphasis on their abstractedness. The fourth paper of this section is 'The Complexity of Performance and the Instruments of Mastery' by Wohlmuth and focuses on the relationship between the embodiments of standards and the critical balance between automatic and creative elements in performance. In Danesi's contribution, a paper entitled 'Meaning Networks and Verbal Communication: A Semiotic Perspective of Human Discourse', there is a claim that verbal discourse involves a 'reflexive loop' between the grammatical and the conceptual domains. This section closes with a paper by Haught that raises the question of 'Can Religion be Reconciled with Science?

'The Challenge of Sustainability' is the theme of Section Four. It is a challenge that three papers address. Fey and Lam start this section with their paper 'The Bridge to Humanity's Future' in which they present the nature of the Ecocosm, its problems and the Paradox. They go on to offer recommendations for finding and implementing a solution to the Paradox as their understanding of the only reasonable chance that humanity has to save it from itself. Odum continues with a contribution entitled 'Energy, Hierarchy and Money'. He shows how the distribution of monetary circulation can be linked to the energy hierarchy, and provides some quantitative insights on the self-organisation of an economy. Shireman embraces the section's theme explicitly in his paper 'Capitalism in the New Millennium: The Challenge of Sustainability'. Shireman gives an illustrative presentation using a range of organisational examples. These organisations have attempted to move from the old industrial machine age into a new age of systems of management and measurement, and of a new set of values that see business and the economy as living systems, sustained by profit, performance and an underlying sense of purpose.

'The Art and Science of Forecasting in the Age of Global Warming' is the heading for a section comprising three papers from the Forecasting symposium that bring historical, practical and reflective slants to forecasting. Linstone opens the section with a paper entitled 'Technological Forecasting and Assessment Tools: the Changing Scene' that sets the context very well for the next two papers. In 'Forecasting the Innovation of Complex Technologies, Kash and Rycroft use three characteristic patterns that have emerged from case studies of the evolution of six complex technologies to develop ideas that will minimise oversight in forecasting. Finally, in 'Making Technology Foresight (and Systems Studies?) Useful', Porter suggests ways that technology forecasting and, by implication, systems analyses, must change to add real value.

The Y2K 'problem' was a 'hot topic' in the final years of the last century, and it was the subject of a special symposium. Three representative papers from the symposium form the sixth section of this volume. Umpleby's paper 'Why the Century Date Change Occurred So Smoothly' answers a range of questions in relation to Y2K from 'what actually happened on January 1st?' to 'why was there so little disruption?' and to questions about the effects of Y2K in countries such as Russia, Italy and USA. Bosch explores which measures used to deal with the Y2K problem are also applicable to the future security of information networks in critical infrastructure by assessing the steps taken by governments, industry and international organisations to deal with the Y2K issue. Her paper is entitled 'The Year 2000 (Y2K) Issue and Implications for Security of International Information Infrastructure'. Müller also tries to take lessons from the Y2K experience. In 'Y2K – "Ten Months After": Two "Double Binds" and a " Double Blindness"' he undertakes a detailed ex-post analysis of the Y2K problem and suggests new ways of risk assessment for contemporary knowledge societies in general.

The penultimate section, Section Seven, takes a look at the challenges and opportunities that might lie ahead of systems thinkers and the systems discipline at large. Contributions from four authors are brought under the heading of 'The Future of Systems Science'. The first paper in this section is authored by Troncale with a focus on natural systems. In his paper 'The Future of the Natural Systems Sciences' he provides an historical and taxonomic panorama of the new sciences of 'systemness' or complexity. Jackson follows with a paper entitled 'Systems Thinking and the Social Sciences'. Jackson addresses the past, present and future of systems sciences by exploring the changing face of social systems theory. Midgley then goes on to convey an optimistic stance in 'Systems Thinking for the 21st Century' by outlining a systems perspective he has been engaged in over the last couple of years (based on Critical Systems Thinking) that is particularly responsive to the conditions that we find ourselves in. The closing paper in this section, from Holmberg, looks ahead to the new millenium in 'OR 2000: Challenges and Prescriptions for the New Millennium'.

The final section of this volume brings together the Presidential Addresses that were delivered at the conference; namely from the Past, Present and Incoming Presidents. In that order, Swanson explores the potential of 'The Systems Movement and ISSS in the New Millennium', while Corning takes a look at 'The Systems Sciences in the Year 3000' and Nelson highlights the contemporary role of ISSS as that of service. In his paper entitled 'Continuing the Traditions of ISSS – Systems Science in the Service of Humanity', Nelson stresses that this is a role he wants to continue to support during his tenure as President.

And so, from the diverse range of topics and differing styles of presentation, we trust that you will be able to follow a personal journey that will enrich your systems work in the future.

CONTENTS

Section Three – Human Systems in the 21st Century

Section Four – The Challenge of Sustainability

Section Five – The Art and Science of Forecasting in the Age of Global Warming

Understanding Complexity

Section One

Theories for Complex Systems

THE SYSTEMIC VIEW OF THE WORLD: ∫ LOGO OF GENERAL SYSTEMS RESEARCH

Anatol Rapoport

Department of Psychology
University of Toronto
Toronto, M5S 3G3 Canada

INTRODUCTION

When I was a freshman, our physics professor, unable to make us feel at ease, told this story. A physics course was described in the catalogue as "non-mathematical". On the first day of the semester, the professor, after saying "Hello", turned his back on the students and, starting in the upper left-hand corner of the wall-wide blackboard, covered it with equations, finishing promptly after fifty minutes at the lower right-hand corner. Someone ventured to point out that the course was listed as "non-mathematical". The professor pointed to the blackboard and said, "You don't see any integrals, do you?"

Indeed, the integral is a fitting symbol of serious mathematics. Being related to making a whole out of parts, it is also a fitting symbol of the systems outlook. It was at one time the logo of the Society for General Systems Research.

As the last survivor of its four founders, I am at times honoured by an invitation to give the opening address of meetings such as this one. This time, I will begin by identifying the themes that initially inspired each of the four of us, and how these themes are reflected in today's scientific and political activities.

GENERAL SYSTEMS RESEARCH

The name General Systems Theory is usually credited to Ludwig von Bertalanffy who called attention to processes apparently involving "equifinality", that is the attainment of a particular final state in spite of external disturbances impinging on the dynamics of a

Understanding Complexity, Edited by Ragsdell and Wilby,
Kluwer Academic/Plenum Publishers, New York, 2001

process. In this way, Bertalanffy refuted the contentions of vitalists that only living organisms are capable of exhibiting "equifinality", that is apparently "purposeful behaviour".

Ralph Gerard was concerned mainly with the integration of biological and social disciplines. He illustrated this integration by a matrix, whose rows represented levels of organisation with which the disciplines were concerned – cell, organism, group, society, international system – and whose columns represented the three main aspects of organisation – structure, behaviour and evolution (being, acting, becoming, as Ralph used to say).

My interests at that time revolved around structural (as distinguished from metaphorical) analogies between apparently disparate phenomena, revealed by mathematical models of these phenomena. For example, the behaviours of a harmonic oscillator on the one hand and an electrical system involving an alternating current, an inductance, a resistance, and a capacitance in series on the other are both describable by the same linear differential equation of the second order. Thereby the inductance of the electrical system is seen as an analogue of the mass of the mechanical one, the resistance of the former of friction of the latter, capacitance of the former an analogue of the inverse rigidity of the later, etc. Similarly, the insight that the same process, namely oxidation, underlies the rusting of a nail, the burning of a candle and the breathing of an animal is an example of integration of knowledge, the process that makes science resemble a cathedral rather than a pile of bricks.

KENNETH BOULDING

The most significant of these approaches (all of which are aspects of integration) was, in my opinion, that of Kenneth Boulding, who distinguished nine levels of system complexity:

1. A system characterised by a structure but no dynamics.
2. A clockwork, having a structure and rigid dynamics insensitive to inputs.
3. A cybernetic system, containing internal feedback mechanisms.
4. An open or self-maintaining system.
5. A genetic societal system, typified by a plant, that is, a collection of differentiated, mutually dependent parts (roots, leaves, seeds).
6. An animal, capable of quasi-teleological (goal-directed) behaviour and beginnings of self-awareness.
7. The human level, where self-awareness develops into reflexion plus communication by symbolic rather than signal language.
8. Human social organisation, capable of planned and conscious co-operation.

And being a devout Christian, Boulding added:

9. Transcendental system, containing ultimates, absolutes, and inescapable unknowables.

In reference to these Boulding adds, "It will be sad day ... when nobody is allowed to ask questions that don't have any answers." (Boulding, 1956).

It is easy to see that this hierarchy reflects sequential expansion of the system concept, in other words, expanding integration. The geosphere has structure only. The hydrosphere and the atmosphere include dynamics. The solar system is essentially a clockwork. Living systems exchanging matter and energy from the environment become self-organising systems; human societies are consciously co-operating systems.

I would make only one substitution. Instead of Boulding's "transcendental system", I would supplement the geosphere, the hydrosphere, the atmosphere, and the biosphere by the noosphere, a term introduced, I believe, by the Russian scientist Vernadsky (1863-1945). It includes all human knowledge, true, doubtful and false. It becomes a self-organising entity to the extent that knowledge becomes integrated into a system, which is the clearest mark of scientific knowledge. It is this integration by generalisations inherent in the scientific method that inspired the use of the integral sign ∫ as the logo of the Society for General Systems Research.

INTEGRATION

I tend to think of the integration principle not only in epistemic but also in political terms. Consider the evolution of human society. Surely, the first co-operative groups were families. In time families joined into clans, clans into tribes, tribes into nations, nations into alliances. The final integration of humanity is already visible on the historical horizon, however in two forms. One portends further integration with all life, enhancing the life expectancy of the supersystem, that is, the biosphere. The other appears to many as a threat, portending the extinction of at least our species within a time span of the order of centuries, perhaps even decades.

Here is why I feel this way. The process of successive integration of human groups was driven primarily by enmity towards other groups. Contacts between separately organised groups were typically collisions, generating conflicts over territory, resources, and later over domain of political control, spheres of influence, etc. So two opposite processes went hand in hand: internal integration, enhancing co-operation and external differentiation, the use of this enhanced co-operation against a common enemy. Strong loyalty toward "us" ignited fierce enmity toward "them".

We observe the same dichotomy in the meaning of "globalisation". One envisages it as free universal co-operation; the other as unimpeded competition. One envisages the global village as a community; the other as the global maker, resembling a Hobbsean war of every one against every one, except that the participants are corporations instead of people, and defeat means going broke instead of getting killed. It is this apparently less lethal version of the Hobbsean struggle for existence that makes it acceptable to those who see competition as the prime mover of progress.

I say "apparently" less lethal, because the lethal consequences of uninhibited competition are either not realised or ignored. The profit motive conspicuously over-rides all others. Maximisation of profit induces unimpeded growth, which suggests an analogy with malignant subsystems of the system in which corporations are imbedded, namely the human race.

The potentially lethal consequences of corporate globalisation have become apparent in the light of the costly expanded noosphere especially in the twentieth century. It is clear that free movement of capital facilitates evasion of any form of control over the activities of corporations guided primarily by maximisation of profits and imperatives of growth. Formidable political obstacles generated by corporate imperatives, bar the way toward conservation of non-renewable resources, toward impeding the degradation of the hydrosphere, the atmosphere, the biosphere. The lure of cheap, politically impotent labour preserves aggravates gross social inequalities on the global scale, thus igniting outbursts of violence. These outbursts block integration of humanity to its supporting systems – the geosphere, the hydrosphere and the atmosphere – and to its supersystem – the biosphere.

CONCLUSION

In conclusion let me cite another remarkable insight of Ken Boulding that has a direct bearing of the systemic view of humanity. Ken distinguishes three modes of social control: threat, trade and love. Threat characterises dictatorships and autocracies: do (or don't do) this or that or else you will be punished. This mode is based on instilling fear in the governed. Trade characterises modern liberal democracies, so this for me, and I'll do that for you. This mode is based on peoples' immediate self interest. Lover characterises the integrated fully co-operative society. People are nurtured, cared for, etc., because they are members. We take care of out children not because they threaten reprisals if we don't, and not because they pay more for our care, but because they are our children. I see the three modes as representing three levels of morality. Trade is more moral than threat, and love is more moral than trade. In the context of relations between sexes, I regard prostitution more moral than rape, and monogamy more moral that prostitution. In the context of politics, I regard democratic capitalism as more moral than totalitarianism, and democratic socialism more moral than capitalism. Later Boulding replaced "love" by "integration", probably feeling that sentimental overtones are out of place in academic discourse. And it is about integration, the key concept of system science, that we have been talking about. I may mention in passing that this word is also semantically related to "integrity".

SYSTEMS SCIENCE FUNDAMENTAL PRINCIPLES

James R. Simms

9405 Elizabeth Court
Fulton, Maryland 20759 USA

INTRODUCTION

Fundamental principles are identified that apply to both nonliving and living systems. These first principles provide the basis for the integration of nonliving and living systems sciences. A brief history of the search for fundamental principles applicable to both nonliving and living systems is provided, which shows that it has been a long search. The fundamental principles are given, followed by the proofs of these principles.

The search for fundamental principles that apply to both nonliving and living systems has been ongoing since Newton's *Principia*. Philosophers of the 18th century attempted to create a science of society that rested on general laws of nature equivalent to the principle of universal gravitation (Westfall, 2000). The concept that living things are similar to other parts of the physical world, differing only in their complexity, was explicitly stated in the early years of the twentieth century by the biologist Ludwig von Bertalanffy. He considered living things to be a part of the natural order, "systems" like atoms and molecules and planetary systems. Systems were described as being made up of a number of interrelated and interdependent parts, but because of the interrelations, the total system became more than the sum of those parts. These ideas led to the development of systems movements that included not only biologists but scientists in other fields as well.

The subjects of systems science are both nonliving and living things. A definition of a system that includes both nonliving and living things found in nature (a concrete, real system) is: a nonrandom accumulation of matter-energy in a region of physical space-time, which is organized into interacting interrelated subsystems or components (Miller, 1978). The basic areas of scientific study of systems are their static and dynamic characteristics. The static characteristics include investigations of the structure and organization of systems. The dynamic characteristics include investigations of systems behaviors when subjected various influences. For example, static investigations of nonliving systems include the

Understanding Complexity, Edited by Ragsdell and Wilby.
Kluwer Academic/Plenum Publishers, New York, 2001

determination of the crystal structure of minerals and static investigations of living systems include the determination of the structure and organization of deoxyribonucleic acid (DNA). Dynamic investigations include determination of the motion behaviors of nonliving matter when subjected to various forces and the metabolic behaviors of living systems when subjected to various food conditions.

The static characteristics of both nonliving and living systems are well developed. The structures and organization of nonliving systems are known with a high degree of accuracy and in great detail. For example, the structure and organization of atoms, molecules, elements, crystals, aggregates, and the earth are well known. There is a vast literature in astronomy, physics, chemistry, geology, and the various engineering disciplines that treat the structure and organization of nonliving systems. A current *Handbook of Chemistry and Physics* is a small example of the vast amount of detail of the structure and organization of nonliving things. The structures and organization of living systems are also known with a high degree of accuracy and in great detail although not to the degree of nonliving systems. There is also a vast literature in biology, anatomy, zoology, sociology, and the social sciences that treat the structures and organization of living systems. Linnaeus' classification system for the identification and naming of living organisms is based on their structure and organization. Miller has classified living systems in terms of their structure and organization.

The dynamic characteristic of nonliving systems, that is, their behaviors, are also well developed. The behaviors of heavenly bodies are well known, as are the relationships between these behaviors and the external influences such as gravity and centrifugal forces. The fundamental principles of astronomy treat these behaviors. The behaviors of physical systems are well known and are explained by Newton's principles. The behaviors of physical systems when subjected to heat are understood in great detail. The behaviors of nonliving systems when influenced by electrical phenomena are well developed. The dynamic characteristics of living systems are known to a lesser degree of precision than those of the nonliving systems.

PRINCIPLES

System science fundamental principles have been developed. These principles are:

(1) The behaviors of systems are observed and quantified by the work done (energy used) in these behaviors.

(2) System behaviors are a function of the energy available to the system.

(3) Each system has an innate capacity to direct energy that is a function of its own unique structure and organization.

(4) There is a direct relationship between a system's behavior and its capacity to direct energy and the energy available to the system.

(5) Living systems are distinguished from nonliving systems by their innate ability to generate information.

(6) There is a direct relationship between a living system's behavior, its capacity to direct energy, the energy available to the system, and its ability to generate information.

DEVELOPMENT OF SYSTEM SCIENCE FUNDAMENTAL PRINCIPLES

Development of system science fundamental principles has evolved from the scientific revolution of the 16th and 17th centuries. There is general agreement that the book that started the scientific revolution was Nicholas Copernicus' *De revolutionibus orbium coelestium (Concerning the Revolutions of the Heavenly Spheres).* Copernicus' book, published in 1543, is taken as the beginning of the scientific revolution because it started new lines of investigation. One of these lines led to Johann Kepler's book *Astronomia nova (A New Astronomy)* published near the beginning of the 17th century. Kepler demonstrated that the behaviors of the planetary system moved in elliptical orbits around the sun as opposed to previous held views that the planets moved in circular motions. Kepler's three laws of planetary motion were the first of the mathematical laws of modern science (Westfall, 2000). These laws relate the motion behaviors of the planetary system to the fundamental measures of length and time.

Another line of investigation evolving from Copernicus' work was conducted by Galileo. In his book *Dialogo sopra i due massimi del mondo (Dialogue on the Two Chief World Systems)*, he investigated the concept of motion behaviors. This book was published in 1632. Before his work, motion was considered to be an expression of a body's nature. Galileo redefined the meaning of motion. He defined motion as a state in which a body finds itself. Galileo extended his investigations into motion in his book *Discorsi intorno a due nuove scienze (Discourses on Two New Sciences)*, 1638. Galileo defined the concept of uniformly accelerated motion, which he identified with free fall; he worked out the relations of velocity, distance, and time in uniformly accelerated motion. A mathematical science of motion behaviors evolved from Galileo's work and became the very heart of the scientific revolution.

A significant concept in the evolution of the systems sciences was presented in the work of René Descartes' series of books, especially *Discours de la methode (Discourse on Method)*, 1637, and *Principia philosohiae (Principles of Philosophy)*, 1644. In essence, he established a relationship between the motion behaviors of matter and the fundamental determinant of these behaviors, which is force. He postulated that the world consists only of particles of matter in motion that lack any spontaneous activity but are inert and move when caused to do so by forces. The above identified investigations of Copernicus, Kepler, Galileo, Descartes, and others provided the foundation upon which Newton developed his world-changing work. Possibly Newton's most important contribution to systems science is the universal gravitation concept and verification that gravity is the major determinant of motion behaviors. However, Newton's concept of energy and the development of this concept is essential to the evolution of the systems science.

It is shown below that the concept of energy (work) is necessary for the fundamental systems science principle that the behaviors of systems are observed and quantified by the work (energy used) in these behaviors.

Behaviors of Systems

Behavior, in its most general meaning, is the way things act. In this context, behavior includes the acts of both living and nonliving systems and is used by astronomers, physicists, chemists, biologists, psychologists, behaviorists, and social scientists to describe the observed activities of the subjects of their particular science. Newton observed the way bodies on earth and planetary bodies behave in a gravitational environment. Newton established mathematical relationships between the behaviors (motions) of bodies and gravitational force. He also established the concept of mechanical work (energy) along with a unit of measure for work. Newton's efforts provided methods of describing observed mechanical behaviors in terms of the energy used in these behaviors and also a means of quantifying the observed behaviors.

Thermal behaviors are a major characteristic of systems. Observed thermal behaviors include the change in size and shape of systems when subjected to thermal forces. The observed behaviors of water when subjected to increased thermal forces include changing from solid to liquid to gas. The observed behaviors of liquids when subjected to increased thermal forces is an increase in the liquid's volume. Liquid thermometers are based on this phenomenon. The amount of behavioral change is a function of the amount of heat energy input to the system. A unit of measure for heat energy, the calorie, is based on the amount of heat required to change the temperature of a specific amount of water by one degree centigrade. There is a direct relationship between the observed thermal behaviors of a system and heat input to the system.

Electrical behaviors are major characteristics of systems. These behaviors include changes in electrical charge and motion of electrons. There are relationships between the electrical behaviors of systems and the energies used in these behaviors. The unit of electrical energy is the watt-second.

A fundamental behavior of living systems is metabolism, the summation of all the biochemical reactions occurring in a living system. These behaviors include the biochemical reactions associated with the synthesis of protein, the replication of DNA, and the glycolysis cycle which converts food into products necessary for life. These biochemical reactions require energy for the conversion from one type of biochemical to another. There is a direct relationship between a biochemical reaction behavior and the work (energy used) done during the reaction. The fundamental unit of energy for biochemical reactions is the energy contained in one molecule of adenosine triphosphate (ATP).

Another fundamental behavior of living systems is reproduction, all living organisms reproduce. These behaviors include the synthesis of protein, replication of DNA, and replication of cells. There is a direct relationship between these replication behaviors and the work done during replication. The fundamental unit of energy for replication behaviors is also ATP.

Animals have an additional fundamental behavior which is muscle contractions. These behaviors include the contraction of an animal's contractile tissue. There is a direct relationship between these contraction behaviors and the work performed during a contraction. The energy used in contractions is in the form of ATP.

The behaviors of nonliving and living systems given above are observed by way of the energy used in the behavior. Observation is fundamental to all science because the scientific method begins with observation. The principle of scientific observation is well established. For example, in his book *Fundamentals of Measurement and Representation of Natural Systems*, Rosen (1978), a mathematical biologist, demonstrates that observation is based on the energies associated with the system being observed. Because energy can be measured and because there is a relationship between behavior and energy, behaviors can be quantified in terms of energy.

Available Energy

Inasmuch as energy is neither created nor destroyed, the energies utilized in the behaviors of systems must be available to these systems for these behaviors to occur. A fundamental doctrine of science is that energy is neither created nor destroyed. This doctrine results in the second principle listed above that systems' behaviors are a function of the energy available to the system. It is concluded that a system's available energy is a fundamental determinant of the system's behavior.

Capacity to Direct Energy

It is observed that the energies in the system behaviors listed above are different from the energies available to the system. The kinetic energy in the motion of a body is different from energy available to the body in the form of gravitational force acting over time. The heat energy of a body is different from solar energy impacting the body. The bonding energies of a biochemical substance is different from the chemical energy ATP that is available to a biochemical reaction. The mechanical energy in a muscle contraction is different from the ATP energy available to the muscle tissue. It is evident based on these observations that the system under observation has directed, that is, operated on, available energy to convert it to the energy in the observed behavior. It is concluded that systems have a capacity to direct energy. Simms (1971), in *A Measure of Knowledge*, gives a mathematical treatment of both nonliving and living systems' capacities to direct energy and demonstrates that this capacity to direct energy is a function of the system's structure and organization. Simms (1999), in *Principles of Quantitative Living Systems Science*, expands the concepts of a living system's capacity to direct energy. He determined that this capacity to direct energy is an innate characteristic of the system, which is a function of the system's structure and organization. It is concluded that a system's capacity to direct energy is a fundamental determinant of the system's behavior.

Relationship Between Behavior, Capacity to Direct Energy, and Available Energy

There is a direct relationship between a system's behavior and its available energy because of the conservation of energy principle. Behaviors occur only when energy has been expended. The energy in a behavior is neither created nor destroyed but must come from the energy available to the system. If energy is not available to a system then the system cannot exhibit behavior. Therefore, there is a direct relationship between a system's behavior and its available energy.

There is also a direct relationship between a system's behavior and its capacity to direct energy. Because the energies in behaviors are the result of a system operating on (directing) energies available to the system, it is clear that the absence of a capacity to direct energy negates any behavior. Also, because a system's capacity to direct energy is a function of its structure and organization, it is clear that the absence of structure and organization means the absence of a system.

Both available energy and a capacity to direct energy are essential for a system to exhibit behavior. It is concluded that a system's behavior is a direct function of both its available energy determinant and it capacity to direct energy determinant. Simms (1983) formalized the mathematical relationship between behavior and these two determinants of behavior.

Distinction Between Nonliving and Living Systems—Information

An ability to generate information distinguishes living systems from nonliving systems. Genetic information in the forms of DNA, ribonucleic acid (RNA), and codons, and biochemical information in the forms of enzymes and hormones are fundamental determinants of living systems' behaviors. Genetic and biochemical information cause the reproduction and metabolic behaviors. Animals have an additional behavior determinant. This determinant is neural information which causes the contractions in contractile tissue. Although living systems have a capacity to direct energy, energy is not directed until caused to do so by information generated by the living system.

Because living systems behaviors do not occur until information causes energy to be directed, there is a direct relationship between a living system's behavior and information. The relationships between living systems' behaviors and information have been developed and units of measure for genetic, biochemical, and neural information have been developed (Simms, 1996 and 1999).

Relationship Between Living Systems' Behaviors, Available Energy, Capacity to Direct Energy, and Information

There is a direct relationship between a living system's behaviors and the fundamental determinants of these behaviors. These determinants are available energy, capacity to direct energy, and information. The absence of any of these three determinants negates the possibility of behaviors of any sort. Fundamental equations have been developed for the relationships between the behaviors of living systems and the determinants of these behaviors (Simms, 1999).

SYSTEMS SCIENCE INTEGRATION

The systems science principles given above provide a basis for integrating the nonliving and living systems sciences. The first four principles apply to both nonliving and living systems. The other two principles extend systems science to include the characteristics unique to living systems. These two living systems principles were developed using the methodologies used in the development of nonliving systems principles. Simms (1999) demonstrated the integration of the nonliving and living systems sciences.

REFERENCES

Miller, J. G., 1978, *Living Systems*, McGraw-Hill, New York.

Rosen, R., 1978, *Fundamentals of Measurement and Representation of Natural Systems*, Elsevier North-Holland, New York.

Simms, J. R., 1971, *A Measure of Knowledge*, Philosophical Library, New York.

Simms, J. R. 1983, Quantification of behavior, *Behavioral Science,* 28:274-283.

Simms, J. R. 1996, Information: its nature, measurement, and measurement units, *Behavioral Science*, 41:89-103.

Simms, J. R. 1999, *Principles of Quantitative Living Systems Science*. Kluwer Academic/Plenum Publishers, New York.

Westfall, R. S., 2000, Introduction in: *Sir Isaac Newton's Mathematical Principles of Natural Philosophy and His System of the World*, The Easton Press, Norwalk.

A COSMOLOGICAL IMPERATIVE: EMERGENCE OF INFORMATION, COMPLEXITY, AND LIFE

Eric J. Chaisson

Wright Center
Tufts University
Medford, MA 02155

INTRODUCTION

We are entering an age of synthesis such as occurs only once every few generations. The scenario of cosmic evolution now grants us an opportunity to systematically and synergistically inquire into the nature of our existence, to mount an integrated effort to build a modern world view that people of all cultures can adopt--a big-bang-to-humankind story that traces generative and developmental change ranging from quark to quasar, from microbe to mind.

Cosmic evolution is the study of change through time--the totality of the many varied changes that have occurred throughout all time and across all space. These are the changes that have produced our Galaxy, our Sun, our Earth, and ourselves. A clear sequence of change has accompanied the arrow of time over the past 12 billion years (Chaisson 1997a). That sequence, as determined from a substantial body of post-Renaissance observations, is galaxies first, then stars, followed by planets, and eventually life forms.

Seven major construction phases in the history of the Universe can be identified: particulate, galactic, stellar, planetary, chemical, biological, and cultural evolution. These are the specialized phases--separated by discontinuities on localized scales--that are responsible for the disciplinary and fragmented fields of reductionistic science. As such, neo-Darwinism is just one, albeit important, subset of a much broader evolutionary scheme encompassing much more than mere life on Earth. In short, what Darwinism does for plants and animals, cosmic evolution aspires to do for all things.

Of central importance, we can now trace a thread of understanding--a loose continuity of sorts--linking the evolution of primal energy into elementary particles, the evolution of those particles into atoms, in turn of those atoms into galaxies and stars, the evolution of

Understanding Complexity, Edited by Ragsdell and Wilby,
Kluwer Academic/Plenum Publishers, New York, 2001

stars into heavy elements, the evolution of those elements into the molecular building blocks of life, of those molecules into life itself, of advanced life forms into intelligence, and of intelligent life into the cultured and technological civilization that we now share. These are the historical phases--much the same as those noted above, but now reidentified from a broader, integrated perspective--that are responsible for the interdisciplinary world view now emerging. The claim here is that, despite the compartmentalization of modern science, evolution knows no disciplinary boundaries (Chaisson 1989; 1997b).

GROWTH OF INFORMATION, COMPLEXITY, LIFE

When matter and radiation were equilibrated in the early Universe (t < 100,000 years), only a single temperature is needed to describe cosmic history; the absence of any thermal gradients dictated (virtually) zero information content, or zero macroscopic order. However, once matter and radiation decoupled, owing strictly to the expansion of the Universe and nothing more, matter became atomic, the gas-energy equilibrium was destroyed, and a single temperature is insufficient to describe the bulk evolution of the cosmos. Since the random motions of the hydrogen and helium atoms failed to keep pace with the rate of general expansion of the atoms away from one another, the matter cooled faster than the radiation (Layzer 1976).

Such a thermal gradient is the patent signature of a heat engine, and it is this ever-widening gradient that enabled matter, in the main, to "build things." The environmental conditions became naturally established to permit a rise in "negentropy" of statistical mechanics (Schroedinger 1944) or in "information content" of the information sciences (Chaisson 1997c). Such non-equilibrium states are suitable, indeed apparently necessary, for the emergence of order, thus we reason that cosmic expansion itself is the prime mover for the gradual construction of a hierarchy of structures throughout the Universe.

The many and real structures observed on all scales throughout the Universe have indeed displayed this sort of increase of order during the course of all time. But rather than specifying that order by using tricky empirical values of negentropy whose measurements are virtually impossible, or by using slippery interpretations of information content whose meaning and connotation are unclear (Marijuan 1996; Matsuno 1996), we prefer to work with a more physically intuitive quantity--in particular, the flow of energy per unit mass, also known as the free energy rate density. This is the quantity that has steadily increased as more intricately ordered structures have emerged throughout cosmic history, indeed dramatically so in relatively recent times (Chaisson 1998; 2001).

CONCLUSION

Cosmic evolution accords well with observations that demonstrate an entire hierarchy of structures to have emerged, in turn, during the history of the Universe: energy, particles, atoms, galaxies, stars, planets, life, intelligence, and culture. As a general trend, we recognize an overall increase in complexity with the inexorable march of time--a distinctly temporalized Cosmic Change of Being, without any notion of progress, purpose or design implied.

With cosmic evolution as an intellectual basis, we can begin to understand the environmental conditions needed for material assemblages to have become increasingly ordered, organized, and complex, especially in the relatively recent past. This rise in order, form, and structure violates no laws of physics, and certainly not those of modern thermodynamics. Nor is the idea of ubiquitous change novel to our contemporary models of Nature writ large. What is new and exciting is the way that frontier, non-equilibrium science now helps us to unify a holistic cosmology wherein life plays an integral role.

REFERENCES

Chaisson, E J., 1989, *Cosmic Dawn: The Origins of Matter and Life*, W.W. Norton, New York.

Chaisson, E. J., 1997a, The cosmic age controversy is overstated, *Science*, 276:1089.

Chaisson, E. J., 1997b, NASA's new science vision, *Science*, v275, p735.

Chaisson, E. J., 1997c, The rise of information in an evolutionary universe, in: *The Quest for a Unified Theory of Information*, W. Hofkirchner, ed., Gordon & Breach, Amsterdam.

Chaisson, E. J., 1998, The cosmic environment for the growth of complexity, *BioSystems*, 46:13.

Chaisson, E. J., 2001, *Cosmic Evolution: The Rise of Complexity in Nature*, Harvard University Press, Cambridge.

Layzer, D., 1976, The arrow of time, *Astrophys. Journal*, 206:559.

Marijuan, P. C., 1996, From computers and quantum physics to cells, nervous systems and societies, *BioSystems*, 38:87.

Matsuno, K., 1996, Internalist stance and the physics of information, *BioSystems*, v38:111.

Schroedinger, E., 1944, *What is Life?*, Cambridge University Press, Cambridge.

THE CO-CREATION HYPOTHESIS

Hector Sabelli

Chicago Center for Creative Development
Chicago, Illinois, 60614, USA.

INTRODUCTION

This article presents a general hypothesis based on empirical studies and mathematical models of creative processes: processes create opposites (bifurcation cascades) and opposites co-create processes and structures. Natural processes are spontaneously creative, displaying pattern transformation, novelty, diversification, and nonrandom complexity. These measurable properties distinguish creative processes (physiological, economic, meteorological series, and bios generated by mathematical models of interacting opposites) from attractive processes (equilibrium, periodicity, chaos), and from conservative processes (mechanical, random). A model of bipolar (positive and negative) feedback generates a development from steady state asymmetric equilibrium to multiple bifurcations, periodicity, chaos, bios, and infinitation. In nature, action (transformation of energy) is the sole constituent of systems; opposite actions coexist; feedback is bipolar; and bifurcation cascades and biotic-like patterns are widely observed. These facts suggest that the interaction of orthogonal actions (similar in one dimension and opposite in another) generates a creative development at all levels of organization. Determined creation represents an alternative to both deterministic and probabilistic models of nature, and to the notion of unavoidable decay to entropic disorder. Co-creation involves both synergy and conflict, integrating one-sided theories of biological evolution, historical development, and socioeconomic processes. Clinical observations and experiments with mathematical models of bipolar feedback show that the more complex patterns emerge when the interacting opposites are fairly equal and of moderate energy. This suggests practical guidelines for family therapy and for social organization.

Understanding Complexity, Edited by Ragsdell and Wilby,
Kluwer Academic/Plenum Publishers, New York, 2001

PROCESSES CREATE SYSTEMS

Systems science was defined by von Bertalanffy (1955) as the development of mathematical-logical principles applicable to many different systems. Scientific principles such as relativity are hypotheses regarding the laws of nature, supported by data, and best formulated as mathematical functional relationships. Mathematical principles describe formal systems; such as lattice, group, and topological theories, which are regarded as the pillars of mathematics (Bourbaki, 1966; Piaget, 1968 for example) and as models for natural processes. Neither hypotheses nor postulates can be proven, but they are refutable by contradictions with data or theoretical inconsistencies. Methodological and heuristic principles such as dialectics derive from reasoning and experience, but are refined through scientific endeavors. The Co-Creation Principle is a hypothesis regarding natural and human processes, and thereby it provides a methodical process approach to scientific, social, and psychotherapeutic practice. As any other hypothesis, it is impossible to prove, but its generality and its unambiguous formulation render it easy to refute, thus satisfying the requirements for scientific status.

Cosmological and biological evolution, social history, and psychological individuation demonstrate that processes are spontaneously creative. Creation involves the formation and transformation of patterns of limited duration (e.g. life), diversification, symmetry-breaking, novelty, and complexification, features that can be measured in empirical data using newly methods developed through the empirical study of creative processes. (These methods are available as a set of computer programs, the Bios Data Analyzer). Diversification is quantified as increase variance with increasing duration of the sample. It provides a classification of processes into three broad categories:

(1) **Conservative processes** (mechanical, random) maintain their phase space volume.
(2) **Attractive processes** converge to equilibrium, periodic, or chaotic attractor.
(3) **Creative processes** diversify, i.e. expand their phase space volume.

Creative processes are generated by the accumulation of opposite changes, generated randomly (statistical noise) or deterministically by recursions of bipolar feedback ("bios", meaning life). Either process can generate the 1/f power spectrum characteristic of many natural processes, but partial autocorrelation indicates deterministic bios rather than statistical noise in the time series of significant natural processes, such as heart rate variation, economic indexes, and meteorological data. In natural systems, feedback is bipolar, rather than purely positive or negative. These empirical facts lead us to reformulate the dialectic notion of complementary opposites (Heraclitus, Lao-tzu) as a scientific principle, i.e. in terms of energy, time, space, information, feedback, and mathematical form.

Co-creation principle: Complementary opposite actions create organization.

Action is defined in physics as the conjoint change of energy and time. Actions are considered to be complementary when they are opposite in one dimension and synergic in another. Creation is defined as a process that generates novel, diverse, and complex organization. Organization is here taken as a system of interactions embodied in a relatively stable material structure as a continuous process that includes both maintenance (homeostasis, self-reproduction) and development from origin to termination.

1. ACTION (OUT OF EQUILIBRIUM ENERGY)

Physical action, the conjoint change of energy and time, is the universal constituent of reality at every level of organization. Everywhere there is movement, nowhere rest or equilibrium. The universe expands, evolves and creates because it is made of action, rather than substance, form or idea. There is nothing simpler than action, because the Plank's quantum, the smallest unit of existence, has the dimensions of action (both energy and time), not of energy alone. Likewise macroscopic processes are made of actions; e.g. cardiac action is the product of the force, duration and frequency of contraction. One and the same stuff underlies the diversity of the universe. This is possible because the "stuff" that makes the universe is action rather than static substance. At the simplest physical level, all forms of energy convert into each other (first law of thermodynamics) and into matter (Einstein's law). Information is not a separate substance; it must be carried by action; ideas exist only as embodied in brains. As physical action is the universal constituent of reality at every level of organization, the infinite variety of the universe is given by differences in form. Action is quantic, made up of discrete units at every level of organization –e.g. Planck quanta, action potentials, atoms. Action is a quantity. Further, action transmits change (transitivity) and carries the asymmetry of time. Thus processes are lattices (sets ordered by asymmetric and transitive relations), one of three pillars of mathematics. Systems are processes, and processes are systems of actions. Systems are made of discontinuous and asymmetric actions, vectors, arrows not point-like elements, so interactions entail synergy, antagonism, or both.

2. NON-EQUILIBRIUM OPPOSITION (INFORMATION)

Each action has an opposite. Unequal opposites carry information, and produce change. Opposition is universal, in space, in time, and in quality. There are opposites in every respect: action and reaction, left and right, positive protons and negative electrons, electromagnetic polarities, complementary DNA strands, anabolism and catabolism, cooperation and competition, certainty and uncertainty. Opposites are distinct but united, complementary, not mutually exclusive. Every physical entity is both a particle and a wave (Bohr's quantum complementarity). Opposite states coexist at the quantum level (superposition principle). Every person is both feminine and masculine, albeit to different degrees. Evolution and decay coexist (enantiodromia), contrary to one-sided views of social progress and of entropic decay. At every level of organization there are fundamental opposites. As processes differentiate (bifurcate), the number and diversity of opposites multiply. Sharing a common origin, opposites are fundamentally similar; diverging from each other, they are essentially different.

Any two actions are similar in some dimensions (at least time) and different in others. Opposites can be synergic (e.g. the opposition of the thumb and the other fingers, opposite sides of a square), antagonistic (subtractive), or both (nonlinear). Non-linear opposites range between these two extremes: they are in part similar and in part different, in part synergic and in part antagonistic. Antagonists need not be inversely related (as one waxes, the other wanes). Naturally occurring opposites can both grow or diminish at the same time, albeit rarely if ever independently. Linear opposites are polar extremes of a continuum; non-linear opposites are located in any relation within a two dimensional plane. The paradigmatic case is orthogonality, interpreted in standard statistics as independence, and here proposed as a model for dialectic complementarity. Simple models for complementary opposition are real and imaginary numbers, and the sine and cosine functions. Validating this trigonometric model for opposition in fundamental physical processes, light waves are constituted by electrical and magnetic fields that are mutually orthogonal (complementarity), bipolar (positive and negative), and continuous (the entire range of opposites rather than only two extremes).

Harmony and conflict always coexist. There is harmony in the tension of opposites, as in the bow and the lyre (Heraclitus). Even homeostatic systems contain mutually antagonistic components; conversely, conflict is not the sole motor of change, and it often destroys more than it creates -- at variance with Darwin, Marx, and Freud. System formation by the assembly of preformed parts, symbiosis, endosymbiosis, and other cooperative processes are fundamental processes in biological evolution (Empedocles, Kropotkin, 1902; Margulis, 1984).

Asymmetry and symmetry are complementary opposites. Fundamental opposites are asymmetric. Massive protons and small electrons make up atoms; women and men make children. Opposition is a fundamental symmetry of nature, and a heuristic principle in physics. Physical entities are postulated, and subsequently discovered, by reasons of symmetry. Fundamental physical processes can be described as a group (a set in which every member has an opposite, its inverse), the second mother structure of mathematics. Each action has a linear and a complementary opposite. The identity element in a group is generated by the sum of 2 inverses, or of a multiplicity of partial opposites. At the quantum levels, uncertainty is a symmetry of opposites. Opposite states coexist at the most fundamental quantum level (superposition), demonstrating the ancient principle of the union of opposites. Above the quantum level, information is an asymmetry or difference between opposites. Equilibrium is rare, because it requires the matching of opposites that are exactly equally intense, exactly linear, and exactly simultaneous. Opposites are unequal, nonlinear, or non-simultaneous. This may be taken as dynamic formulation of the principle of no-contradiction. Opposites are thus separated in space, time or quality (Aristotle's local formulation of the logical principle of no-contradiction), yet they coexist as components of every process (global principle of dialectic contradiction). Even when quantum events are Schrödinger's famous cat is either dead or alive, irrespective of whether or not the box has been opened and an observation has been made, even when the event that may trigger the cat's death is probabilistic, and thereby unpredictable until an observation is made. At the process level, every living organism eventually dies, and while alive it is continually renovating its cell, literally living and dying at the same time. Process logic thus formulates the union of opposites as **quantum superposition, local logical no contradiction, and dialectic global contradiction**.

Being universal, opposition manifests as multiple patterns of action: flux (thermal, quantic), rotation (from electron spin to galaxies), bifurcation, and periodic and aperiodic oscillations. Being universal, opposition starts with twoness, and multiplies by repetition, tending towards an infinite number (circle) of oppositions. Opposition cannot be reduced to two values in either logic or sociology. A cascade of bifurcations, a repetition of opposition, creates 2^N opposites. Iterating negation allows one to construct a mathematical dialectic logic in which true and false, certainty and uncertainty can coexist in various degrees, rather than being mutually exclusive, and in fact generate each other. Action evokes reaction in mechanics and in human processes. Growth depletes the environment. False hypothesis pave the way towards greater insight.

3. CO-CREATION (STRUCTURE AND NON EQUILIBRIUM ORGANIZATION)

Processes create opposites (bifurcation cascades) and opposites co-create processes and structures. Linear and equal opposites create transient equilibrium. Nonlinear (complementary) opposites create, transform, and destroy episodic, novel, diverse and complex (higher dimensional) patterns. Triadic equilibrium generates stable material structure. Multiple interactions generate evolving higher dimensional organization. Two sexes procreate a new individual; oppositely charged particles combine to form atoms; subsystems differentiate within a system; cosmological symmetry-breakings generate opposites; processes engage in mutual feedback. Self-organization is a co-creation resulting from interactions within a system and between the system and its environment. As energy is conserved, formation always is a transformation. Transformation is the subject matter of topology, the third pillar of mathematics. A simple topological model of growth limited by the consequent depletion of the environment, the logistic recursion, shows bifurcation cascades from an initial steady state to 2, 4, 8, ...2^N periods and aperiodic chaos; one opposition generates many others, and thereby a sequence of patterns. Many natural processes display bifurcation cascades.

Bipolar Feedback Generates Bios

The ceaseless feedback generated by the interaction of a system with its environment is a simple and paradigmatic case of co-creation. As natural environments are enormously diverse, feedback usually is bipolar: the output of any system is synergic to some processes in the environment, and opposed to others, and in turn systems continually receive synergic and antagonistic inputs. Likewise, a change in the market elicits both the buying and selling of stock, and a political statement attracts some and repels others. Bipolar feedback is creative. This is illustrated by the process equation (figures 1 and 2)

$$A_{t+1} = A_t + g_t * \sin A_t,$$

in which A_t models action, that is additive and changes in time; the trigonometric function, ranging from 1 to –1, models bipolar feedback; and the feedback gain g varies with time t. As the intensity of the feedback increases, the recursion generates convergence to a steady state, a cascade of bifurcations to 2, 4, 8, ...2^N periods, chaos, and bios. This represents a progression in complexity. Periods are stable. Chaos is sensitive to initial conditions, but the overall pattern remains uniform in time. Bios shows greater morphological diversity and sensitivity to initial conditions than chaos, and expands in amplitude and variance.

Bios is a newly found type of organization composed of episodic and novel patterns, with a beginning, a development and an end. Biotic series show novelty (less recurrence, i.e. more variable, than their randomized copies), diversification (increase variance with increased duration), and non-random complexity. Diversity, novelty, complexity, and episodic patterning are hallmarks of creativity, found in natural processes and in mathematical bios, but absent in chaos. Bios resembles statistical noise generated by the addition of random changes, and may display 1/f power spectrum like pink noise, or a $1/f^2$ power spectrum like random walk (Brownian motion). Bios also includes transient periodicities and linear transitions, features that, together with sensitivity to initial conditions and partial autocorrelation, differentiate deterministic bios from statistical noise. Deterministic bios is to statistical noise as deterministic chaos is to random.

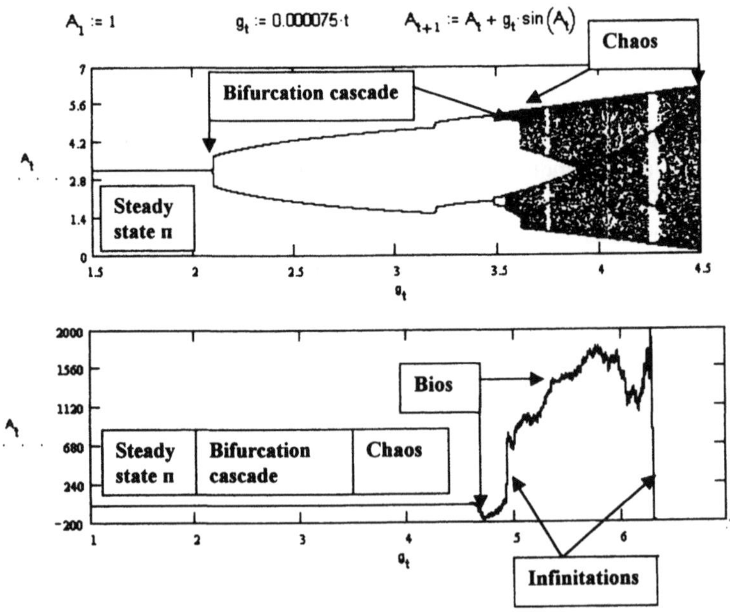

Figure 1: Recursion of bipolar feedback with increasing gain g generates steady state (_), cascade of bifurcations, chaos, bios, and infinitation. The time series is presented with different scales to display stages of markedly different in amplitude.

Bios is found in empirical time series portraying meteorological, biological, and economic processes. This widespread occurrence of biotic patterns suggests that bios may be the canonical form of natural processes hereto suspected to be random or chaotic. (Biotic series have been regarded up to now as chaotic or random because there were no methods to distinguish them). As only bipolar feedback also generates bios, bipolar feedback may be a significant component of natural processes. These observations support the hypothesis that the interaction between synergic and antagonistic complementary opposites in natural processes may be a major factor for creative evolution and history. The analysis of time series of prices, currencies, and indexes, show that economic processes have biotic patterns, suggesting that the interaction of supply and demand does not lead to equilibrium but instead represent a bipolar feedback that generates novelty and creativity. If economic processes are creative, undesirable phenomena such as excessive poverty are not "the unavoidable consequences of the laws of economics", but a matter of choice that can be changed.

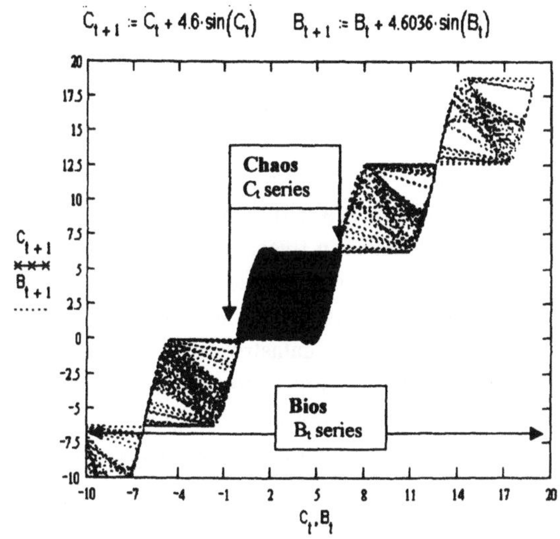

Figure 2: Cobweb diagram of bipolar feedback recursions. Ct generates a chaotic trajectory that remains inside one cycle of the sine wave generator (attractor as stable trajectory). B_t, with slightly higher gain, generates a biotic trajectory that expands positively and negatively visiting many cycles (expansive process). The aperiodic biotic trajectory reconstructs the periodic sine wave that generates it. The sine wave is both the generator and the final attractor (in a different meaning of the term) of the biotic process.

Tridimensional Balance Generates Stability (Structure, Matter)

The three-way split of light into primary colors in the retina signals the transition from physics to biology, points to trifurcation in the creation of complexity. Triadic organization is fundamental: tridimensional space, three colors of quarks that make hadrons, three primary colors, and many conceptual systems. Period 3 implies chaos and an infinite sequence of periodicities and infinitations (Sarkovskii's theorem). Tridimensional balance provides the stability of structure that characterizes matter. In cosmological evolution, there is a net conversion from energy into matter and information I. The symmetric relation "=" in Einstein's law $E = m * c^2$ represents two opposite but unequal actions. As the speed of light c represents the maximum rate of energetic interactions, and communication involves two simultaneous actions, c^2 represents the maximum rate of communication. Interpreting c^2 as information in Einstein's law, energy E becomes matter M and information I:

$$E \rightleftarrows M * I.$$

In a similar manner, interactions may generate higher dimensional organization, such as molecules, organisms, and social systems. These higher forms of organization evolve, transforming patterns, generating novelty, diversity, and complexity. Apparently, three dimensions offer the possibility of a stable balance that does not obtain in lesser or higher dimensions. Notably, the time series generated by bipolar feedback is bidimensional up to the chaotic phase and becomes tridimensional when it reaches the biotic phase. In the same manner, morphological and dimensional complexity may be generated in nature.

Catastrophes: The Sum and Difference of Opposites

Thom's catastrophes illustrate the creation of a tridimensional form by the interaction of orthogonal opposites. A catastrophe is a nonlinear discontinuous change from one pole to its opposite. The study of change between rejection and choice in a clinical setting made us aware that the two parameters that control fold catastrophes are functions of the interacting opposite motivations. The difference between opposites is the information that determines the choice (a, asymmetric parameter of the catastrophe). When the opposite motivations are equal, their intensity provides the bifurcating factor (b) of the catastrophe: ambivalence and back and forth switching occurs if attraction and repulsion are strong, and indifferent neutrality if they are weak. Generalizing, in the interaction between processes, the sum of complementary opposites A and anti-A is the energy E of the system S, and the difference between them provides the information I:

$$\mathbf{b = E(S) = f\,[A(S)\ and\ anti\text{-}A(S)]}$$
$$\mathbf{a = I(S) = g\,[A(S) - anti\text{-}A(S)]}$$

Conversely, increasing the energy of a system increases both opposites; e.g. high energy binds and split atoms, energetic persons make stronger bonds and trigger stormier conflicts. In recursions of bipolar feedback, pattern becomes more diverse and complex as the gain of feedback increases, but at extreme intensity there is a simple linear increase (infinitation). The complexity of pattern decreases when one biases (positively or negatively) the co-creating opposites (sine wave) from exact symmetry about 0. When the sine wave is only positive or only negative, only simple steady states or infinitation obtain. Plotting the complexity of pattern as a function of gain (energy) and asymmetry, one obtains a diagram equivalent to the graph of choices as a function of the sum and the difference of opposite motivations. Both the clinical and the mathematical experiments show that the more complex patterns emerge when the interacting opposites are fairly equal and of moderate energy. This is consonant with clinical experience and suggests practical guidelines for family therapy and for social organization (e.g. marital equality, political opposition rather than consensus, pluricentral management, competition rather than monopoly).

Creative Development

The sequence of bifurcations from one initial state to a pair of opposites, multiple oppositions (2^n periods), chaos, bios and eventual termination roughly models natural processes such as the cosmological and embryological developments. Systems are processes, continually generating diverse, novel and complex patterns of action, and developing from origin to death in the long run. Only in a middle temporal range can systems be regarded as static (homeostatic) and continually self-reproducing (autopoiesis). Systems such as stars or organisms, develop along generally predetermined patterns, but also generate novelty and individuation. Evolution may thus spring from a development determined by universal mathematical laws leading to creative stages, such as bios, that generate novelty, complexity and individuality.

As physical action is the sole constituent of reality at every level of organization, processes and the systems they generate are determined by their organization, i.e. their form and their dimensionality. Novel forms and greater dimensionality define the chemical, biological, social and psychological levels in which physical action is organized. Each level includes the dimensions of the previous levels, and adds new ones. Evolution may thus be regarded as a process of dimensiogenesis. According to process theory, one-dimensional action, bidimensional opposition, and tridimensional space are the material embodiments of primary mathematical forms. A process is a lattice of actions (change in time as ordering relation), a group of rotations (including an inverse or opposite for every action), and a topological transformation of matter in multidimensional space. This hypothesis formulates the ancient notion of cosmic forms (Pythagoras, Lao-tzu, Pasteur, Gödel, Jung) as actions. As such, they are **generators**, they are embodied in the present, and they give form to the entire process. Acting, opposing (informing) and organizing are three inseparable aspects of each process. Action plus opposition (sine wave) combined in a feedback arrangement appear sufficient to generate universal patterns (equilibrium, periodicity, chaos, bios, infinitation) present at all levels of organization. While the biotic trajectory continues to diversify indefinitely, it also recreates the sine wave form that generated it. The generator-attractor can be seen in the cobweb plot (figure 2).

Feigenbaum's cascade of bifurcations suggests how symmetry-breakings may generate diversity in nature. That bipolar feedback generates novelty, complexity and diversity suggests that development is both determined and creative, offering an alternative to the notion of random origin and evolution of the universe and of life, as well as to deterministic models (equilibrium, chaos) current in biology and economics. The process equation represents a concrete instance of a simple sine wave (linear time + circular opposition) generating complex patterns. The bifurcation diagram of the process equation, from 1 action to 2^N values (information), to chaos (determined but unpredictable), to structure-like period 3, to bios, thus suggests to us a scheme for cosmic development as the generation of novel, diverse and complex (higher dimensional) forms from universal, low dimensional, necessary mathematical forms:

Universal forms: 1D Action → 2D Information → 3D Matter →
Created forms: many D Life → high D Mind → ∞D Attractor

SCIENTIFIC IMPLICATIONS AND PRACTICAL APPLICATIONS

As systems contain opposites, and interact with their environment, bipolar feedback may be expected to occur at all levels of organization. Thus systems are autodynamic and creative, not only homeostatic or autopoietic. Creation implies the diversification of systems, at variance with equilibrium models of health and economics, and with homeostatic models of family, social systems, and the planet (Gaia).

Co-creation represents an alternative to the notion that conflict drives biological evolution (Darwinism), economic progress (standard economics), and human history (Marxism), a belief that can undoubtedly foster conflict, but it also contradicts the portrait of systems as integrated and non-conflictual, a view that promotes acceptance of existing but not always desirable systems.

A determined creation of complexity likewise contradicts the notion of unavoidable decay to entropic disorder. If creation is the spontaneous result of interaction, the universe evolves towards an infinitely complex attractor. Empirical evidence demonstrates evolution, and a number of ways to accommodate the second law of thermodynamics have been suggested. In recursions of bipolar feedback, statistical entropy increases from steady state to periodicity, chaos, bios and infinitation. Infinitation, an ordered process, has higher statistical entropy than uniform random distributions with the same range. This is understandable because statistical entropy measures symmetry and diversity, not disorder.

Restating the **co-creation hypothesis**: The universe spontaneously evolves from necessary mathematical relations (lattice order, group inverse, topological continuity) embodied as action, opposition, and tridimensional space. They create generic patterns (steady states, periodic cycles, chaos, and bios) at all levels of organization. In a similar manner, they may generate relatively stable structures, such as matter, and evolving higher dimensional organization (chemical, biological, social and psychological), and beyond towards an attractor of infinite complexity.

This concept guides the development of methods that differentiate creative from attractive processes, e.g. bios from chaos. Co-creation also provides strategies for management, psychotherapy, and social action.

Acknowledgments

This research is supported by the Society for the Advancement for Clinical Philosophy.

REFERENCES

In this brief article, only basic references to process theory are listed. The reader may also find supporting data and references in the Proceedings of the ISSS (1990-2000).

Bertalanffy, L. von, 1955, General systems theory, *Main Currents in Modern Thought*, 71, 75-83

Bourbaki, N., 1966, *Elements of Mathematics*, Addison-Wesley, Reading, MA

Kauffman, L., and Sabelli, H., 1998, The process equation, *Cybernetics and Systems* 29: 345-362.

Kropotkin, P.A.K., 1902, *Mutual Aid: A Factor of Evolution*, Heinemann, London

Margulis, L., 1984, *Early Life: How Cells First Evolved*, Jones and Bartlett, Boston, MA

Piaget, J., 1968, *Structuralism*, Routledge and Kegan Paul, New York

Sabelli, H., 1989, *Union of Opposites: A Comprehensive Theory of Natural and Human Processes*. Brunswick Publishing, Lawrenceville.

Sabelli, H., 1998, Union of opposites: from Taoism to process theory, *Systems Research* 15: 429-441.

Sabelli, H., 1999, Process theory: mathematical formulation, experimental method, and clinical and social application, in: *Toward New Paradigm of System Science*, P. Y. Rhee, ed., Seoul National University Press, Seoul.

Sabelli, H., (in press). Novelty, a measure of creative organization in natural and mathematical time series, *Nonlinear Dynamics, Psychology, and Life Sciences*.

Sabelli, H., and Carlson-Sabelli, L., 1995, Sociodynamics: the application of process methods to the social sciences, in: *Chaos Theory and Society*, A. Albert, ed., I.O.S.Press, Amsterdam, Holland.

Sabelli, H., Carlson-Sabelli, L., Patel, M., and Sugerman, A., 1997, Dynamics and psychodynamics: process foundations of psychology, *Journal of Mind and Behavior* 18: 305-334.

Sabelli, H., and Kauffman, L., 1999, The process equation: formulating and testing the process theory of systems, *Cybernetics and Systems* 30: 261-294.

WHAT IS INFODYNAMICS?

Stanley N. Salthe

City University of New York
Binghamton University
HCR 86, Box 228 Laurel Bank Avenue
Deposit, NY 13754, USA

INTRODUCTION

Infodynamics is the study of the accumulation of informational constraints during system development. This is driven by the Second Law of Thermodynamics because (a) macroscopic configurations provide increased opportunities for enhanced external production of physical entropy, and (b) energetic exchanges tend to give rise to mutated forms, increasing the number of constraints in a system, as well as providing more opportunities for entropy production. So, increased amounts of information (I) generate potential increases in informational entropy (H) as well as facilitating physical entropy (S) production. We can identify two kinds of system information -- that which characterizes a system as the kind of system it is (enformation), also informing its constitutive changes, and information acquired as a result of a system's experiences in the world (historical information, or intropy). Intropic constraints can become integrated with system enformation, the final form of which emerges gradually. The emergence of system developmental reorganization requires information, not only from the system itself, but from its environment, as boundary conditions. This raises the question of structural attractors influencing system development as final causes.

INFORMATIONAL ENTROPY

My definition of infodynamics (meaning 'information dynamics') is: Study of the accumulation of informational constraints, as degrees of freedom, during the development of dissipative structures.

Understanding Complexity, Edited by Ragsdell and Wilby,
Kluwer Academic/Plenum Publishers, New York, 2001

I equate information itself with the fixation of a constraint bearing upon a system from any source -- that is, as a reduction of degrees of freedom in available informational constraints. The developmental sequence is (a) nothing, (b) the appearance of informational constraints (symmetries or available degrees of freedom, (c) the generation of information itself by way of the breaking of symmetries and reduction of degrees of freedom.

In this paper I will use Shannon's H -- that is, information capacity or informational entropy (representing variety and uncertainty) -- as a vicar for information itself since there is as yet no general definition of information independent of H that works for all kinds of systems -- *and* because H has been used in infodynamics to obtain the results I will mostly be talking about, which have to do with an expanded understanding of the role of the Second Law of Thermodynamics. I believe that it is time now to begin branching out to other constructions of information, but here I am only reviewing.

Other things being equal, the more informational constraints that are present, the more behavioral variety could be generated by reducing the degrees of freedom they represent in different ways -- e.g., the more genes there are, the more cell types, or species with different behaviors, there could be; the more species, the more ecological interactions -- these cases are isomorphic with the statement that the larger a lexicon is, the more message uncertainty is possible. Even though fixation of informational constraints results in routinely reigning in possible behaviors, periodic and pathological situations can release some of them again, so that information *is* potential informational entropy. That is, maximum H is proportional to, or set by, the number of informational constraints bearing upon a system. In the context of H, information itself can be defined as those fixed constraints that keep a system away from informational equilibrium, where every configuration it could potentially assume, given the kind of system it is, would be accessible over the natural period of that system's behavior, reflecting all of the system's informational constraints (Figure 1).

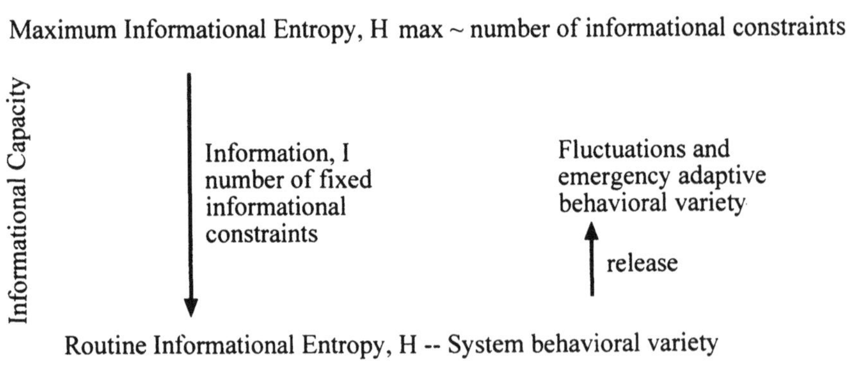

Figure 1. Information as fixed informational constraints.

Since H is a statistic:

(a) The natural period of a system's behavior (based on a system's scale, and analogous to relaxation time in equilibrating systems) is required to be known so as to obtain conditions for observing permutations of behavioral variety. *And*

(b) In order for the distribution to sum to unity (normalizability), H must be based on an external observer's categories. That is, only behaviors of consequence to the observer are identified as possible. Infodynamics recognizes the need to identify a system of interpretance that would be using, or detecting, the information supposedly held in various configurations. For the most part, so far, interpretation has been held to be the province of an external observer. Internalist approaches, including Peircean semiotics, are only now getting under way, and play only a small role in this presentation. I will be referring mostly to the behavioral variety detectable, or, indeed, constructed, by an external observer.

Fixed informational constraints (measured as a deficit from maximum behavioral variety) not only prevent a system from falling apart into any old behaviors whatever, thus keeping it (in a sense) from being any other kind of system, but also enable a system's behavior. A totally unconstrained situation can give rise to no discernible behavior (defined as categories of repeated characteristic actions -- the observer again!). Furthermore, such an unconstrained situation would not express a comparatively large informational entropy either because it would be completely vague, and so would not be a system at all. Only a constrained situation can be conceived as *having* a potential informational equilibrium (including in the permutations many pathological configurations), which its routine behavioral variety will hold it away from. Again, only a constrained situation could be conceived, in fact, to *be* a system.

SYSTEM CHANGE

I define development operationally as 'predictable directional change' (Salthe, 1993), usually implying change that is constitutive of the kind of system involved. During its development, a material system exchanges a primal vagueness for degrees of freedom, and these for increasingly definite embodiment. So it changes generally in such a way that it can inhabit increasing numbers of distinct configurations -- that is, its development generates H because during growth and differentiation it is both acquiring new informational constraints and reducing some of them.

Quantitative Change

Maximum behavioral variety always increases in a developing system -- simply by way of system growth, which adds informational constraints (Figure 2). The expected shape of the increase against age would be roughly hyperbolic, reflecting material constraints against unlimited amounts of configurations -- a "law" of diminishing returns, if you like. (This does not take into account alternate reading frames, but that would require alternate reading configurations.) The amount of fixed information is entrained to a similar shape (the distance between maximum and routine H in figure 1). On the asymptote the system becomes senescent from information overload, imposed, again, by limited configurational resources. The main result of this overload is system inflexibility -- some of the new information just goes to reinforce system habits, imposing rigidity and restricting the variety requisite for adaptability. This eliminates, especially, configurations that might be adaptive to rarer perturbations. But some of the new information imposes lags in routine information flow by insinuating alternate pathways which interfere with existing functions, slowing internal communications and behavioral reactions.

Empirically, H increases along with physical entropy production, because, as understood in Big Bang cosmology (Layzer, 1976; Frautschi, 1982; Landsberg, 1984), the appearance and subsequent clumping of matter can be seen to result from an increasingly intense search for thermodynamic equilibrium. A necessary increase in H can be derived as a result of the Second Law of Thermodynamics (Swenson, 1997; Schneider and Kay, 1994), as follows. Material configurations (the results of fixed informational constraints) facilitate energy gradient degradation and, therefore, external physical entropy production, by mediating a switch from frictional conduction to rapid convection. So, variety of configurations -- e.g., biodiversity (connected as that is to ecological niche diversity) serves the Second Law. The more kinds of species there are (as brought about by dispersion, competitive exclusion and character displacement), the larger the variety of energy gradients being degraded, and consequently, the faster would physical entropy be produced in a given locale.

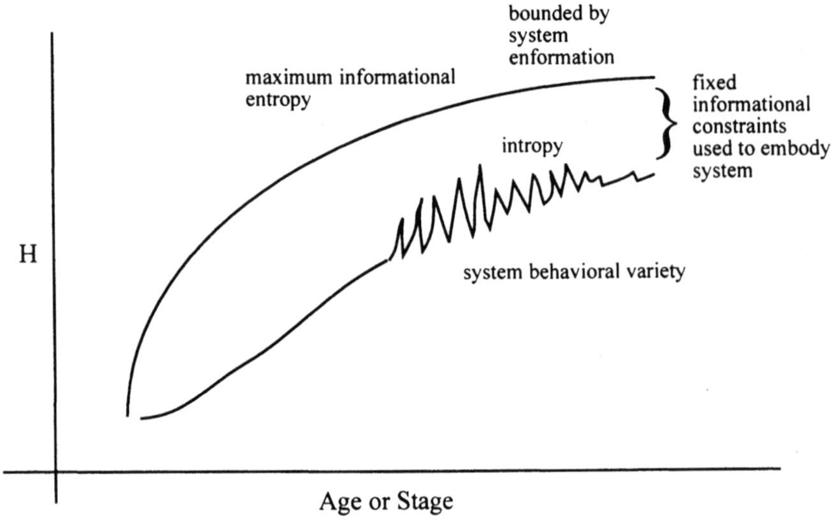

Figure 2 Generalized information dynamics.

Developmental increases in functional routine behaviors in organisms also correlate with increased external physical entropy production, and, both here (Zotin, 1972) and in ecosystems (Schneider, 1988) increases in such variables (e.g., heat transfer) slow down in a similar pattern after senescence sets in. So, roughly, the more fixed informational constraints, the more entropy may be produced. And so development may also be viewed as being driven by the Second Law, until the point of information overload, after which, nevertheless, increases in H continue to occur, but at reduced rates. Summing up this argument -- increases in informational constraints, as in development, allow for a greater variety of system behaviors (H) enabled by more fixed constraints, which behaviors allow more opportunities for physical entropy production. Other things being equal, the more informational constraints, the more H *and* the more physical entropy being generated. Therefore the Second Law can be seen to indirectly elicit increases in H as well. Physical entropy and informational entropy increase together in developing systems.

An infodynamic postulate derived from these results is: informational entropy, or information carrying capacity, must increase in expanding or growing systems.

Taking the internalist position (e.g., Salthe, in press) for a moment, there is also reason to believe that, for any behaving material system, H, as confronted externally from its environment, must tend to increase, because any of a system's behaviors that succeed in reducing local uncertainty *for a system* will cause an external chain of events ultimately giving rise to increases in the uncertainty that system may meet later from its more global surroundings (Salthe, 1990). The more you find out, the more you have disturbed your environment! The result will be that the environmental uncertainty a system must deal with will never decrease globally, delivering, from an internalist point of view, a purely informational model of the Second Law.

We might note here that, from the point of view of a system rather than an external observer, the information actually used by the system (referential information, Kampis, 1991) is most likely not the same as that observed externally. In this context we could note Bateson's (1979) definition of information as "a difference that makes a difference". This connects with information as carrying potential surprise (Dretske, 1981). What this means here is that only informational constraints fixed differently from a system's own version will potentially have meaning for it. That means, among other things, that the information used by a system would not be that which characterizes it *as* a system (John Collier's, 1990 concept of *enformation*), but certainly must include some of the information it has acquired as a result of its experiences (Collier's *intropy*). My interpretation of enformation is that it is for the most part derived from environmental affordances, which generate a system's informational constraints, or, what is the same thing, its enformation. A system is a *part* of its environment, which has made a place for it among its own configurations, and allows it to grow into the world. As it grows a system acquires further, intropic, information as marks of its adventures, when some of its still open degrees of freedom get reduced, increasing its internal information tally. Some of this acquired information could be incorporated into its enformation by a system's self-organization as it individuates, diverging in various ways from the standard or average for its type, which is set originally by its environment.

We might dwell for a moment on the sense of information as a difference or distinction. For example, genetic information is always associated with some phenotypic difference -- no difference, no gene (even if you've got DNA!). This raises the theoretical problem of similarity. In the informational world of distinctions, similarity is just the absence of difference. This covertly assumes that all of the world was at one time a uniform whole and has subsequently differentiated. In general, and from the point of view of Natural Philosophy (which is now my own discourse), this would not be problematic, I think, but there are some troubling facts -- for example, the reality of convergent evolution in biological systems, and of similarities in the vegetations of some far flung biomes. These appear to be similarities *not* devolving directly from common ancestral systems. Some of us have postulated structural attractors to explain such phenomena. In this case the material commonality might go back before the origin of life, to the establishment of laws of matter -- those that dictate, for example, the forms of trees and vortices. Structures would be partially fixed, large scale informational constraints, partially reduced early in the history of the universe, with rhythmic, but unstable realizations which depend upon other partially reduced situations to supply missing information. They act as relatively weak forces at every scale, such that, when a material system appears that supplies enough further fixations of associated constraints, the form of the structure (say, the tree form or the vortex) is accessed, because the degrees of freedom have been reduced so far as make the form emerge from immanence to actuality. Structures are immanent forms. In the immediate biological cases at hand, this would mean that there would be final causes (the pull of attractors) at work in the material world at any given moment now.

Qualitative Change

So, information can be divided into that which contributes to system individuation or evolution -- historically acquired, intropic information -- and that which characterizes the system as that kind of system -- its enformation, which may be taken as responsible for the constitutive changes a system undergoes as it develops, bending intropic acquisitions to the service of these changes as well. I define evolution, or individuation, as the irreversible accumulation of historical information (Salthe, 1993), or intropy -- that is to say, as the acquisition of informational constraint reductions not strongly entailed by previous ones, carrying a deal of surprisal (Dretske, 1981).

Historical information feeds upon itself quantitatively, by way of mutations (e.g., Brooks and Wiley, 1988), which generate more of it as a result of the frictional materiality of the world, and which can be taken, once again, to be driven by the Second Law of Thermodynamics, in the sense that any informational configuration will tend to be eroded or modified by energy transactions in its vicinity. The more intropic information there is, the more variety could be produced from its mutations, generating more possible deviations from routine behavior. But, as a system senesces, its own material inertia will enlist more and more of these mutations into its by now overdetermined behavioral repertory as tributaries. There is, therefore, a widening gap between the maximum variety possible for system behavior -- Hmax, based on possible permutations of a system's configurations -- and the average actual variety demonstrated by a system as it continues to endure (Brooks and Wiley, 1988).

System changes devolving from its enformation, seen externally as predictable developmental changes, depend to a large extent upon an observer's categories, as expressed in normal tables of development, cut up into developmental stages. However, with enformation taken basically as an informational aspect of a system's environment, developmental changes can also be interpreted as constitutive because, in order for a system to persist, its environment must continue to elicit and entrain it by way of its own larger scale, and longer acting, rhythmic behaviors. Semiotically (e.g., Salthe, 1998), developmental changes could be viewed as a system's interpretation of its environment's affordances, which, when particular ones of these are widespread and persist, give rise to classes of behavior with repeated instances, as is prominent in biology. Class properties, being repeated regularities, lead to observations of predictable changes. Here we encounter similarities again, this time interpreted as being imposed by relatively unchanging larger scale environmental affordances.

These considerations can lead us to the concept of emergence -- the appearance of changes that are *not* predictable from an observer's account of the organization of a previous stage (even when predictable empirically). Examples would be entrainment of individuals into orderly group behaviors (as in tissues of cells, flocks of birds, schools of fishes and armies of men), or gastrulation out of a blastula, or physical phase changes, as well as the properties of chemical compounds compared to those of their uncombined constituents. In all these kinds of changes in behavior -- developmental because predictable empirically -- we find new forms appearing, by way of cohesion, at a higher scalar level than the previous system. Cohesion requires, not only system enformational constraints, but also other kinds of environmental information, such as boundary conditions -- (viz. the basic triadic system of Salthe, 1985). New behavioral possibilities seem to open up as a system changes scale. This is fundamentally why emergent phenomena are not generally derivable from theories explaining previous stages or situations.

So, in my view, enformational constraints alone would not be sufficient to mediate the emergence of major reorganizations of a developing system. Yet such reorganizations appear to be constitutive of many such systems -- as with caterpillars and butterflies. One way to explain this might be to suppose that system growth creates material situations whereby previously covert structural attractors come to accessed, which, in the context of the fixed enformative constraints constitutive of the system, which supply the additional information needed to realize the structure's form, spark system reorganization. And, of course, in the new organization, previous enformative constraints could easily have very different material entailments.

So we find that the concept of environmental informational constraints leads once again to the notion of structural attractors of material systems, which might be postulated as explaining, in principle, facts like convergent evolution -- or any other similarities not derived by replication from identical, relatively recent ancestral situations. Have we, again, faced with convergent evolutionary phenomena, raised the possibility of final causality in nature?

Acknowledgment

I thank Mishtu Banerjee, John Collier and Edwina Taborsky for stimulating comments.

REFERENCES

Bateson, G., 1979, *Mind and Nature: A Necessary Unity,* E. P. Dutton, New York.

Brooks, D. R., and Wiley, E. O., 1988, *Evolution As Entropy: Toward a Unified Theory of Biology,* University of Chicago Press, Chicago.

Collier, J. D., 1990, Intrinsic information in: *Information, Language, and Cognition,* P.P. Hanson, ed., University of British Columbia Press, Vancouver.

Conrad, M., 1983, *Adaptability: The Significance of Variability from Molecule to Ecosystem,* Plenum Press, New York.

Dretske, F. I., 1981, *Knowledge and the Flow of Information,* MIT Press, Cambridge, MA.

Frautschi, S., 1982, Entropy as an expanding universe, *Science* 217: 593-599.

Kampis, G., 1991, *Self-modifying Systems in Biology and Cognitive Science: A New Framework for Dynamics, Information, and Complexity,* Pergamon Press, Oxford.

Landsberg, P. T., 1984, Can 'entropy' and 'order' increase together?, *Physics Letters* 102A: 171-173.

Layzer, D., 1976, The arrow of time, *Scientific American 233: 56-69.*

Salthe, S. N., 1985, *Evolving Hierarchical Systems: Their Structure and Representation,* Columbia University Press, New York.

Salthe, S. N., 1990, Sketch of a logical demonstration that the global information capacity of a macroscopic system must behave entropically when viewed internally, *Journal of Ideas* 1: 51-56.

Salthe, S. N., 1993, *Development and Evolution: Complexity and Change in Biology,* MIT Press, Cambridge, MA

Salthe, S. N., 1998, Naturalizing semiotics, *Semiotica* 120: 381-394.

Salthe, S. N., in press, Theoretical biology as an anticipatory text: the relevance of Uexküll to current issues in evolutionary systems. *Semiotica.*

Schneider, E. D., 1988, Thermodynamics, ecological succession and natural selection: a common thread, in: *Entropy, Information, and Evolution: New Perspectives on Physical and Biological Evolution,* B.H.Weber, D.J. Depew and J.D. Smith, (eds.), MIT Press, Cambridge, MA.

Schneider, E. D., and Kay, J. J., 1994, Life as a manifestation of the Second Law of Thermodynamics, *Mathematics, Computation and Modeling* 19: 25-48.

Swenson, R., 1997, Autocatakinetics, evolution, and the law of maximum entropy production: a principled foundation toward the study of human ecology, *Advances in Human Ecology* 6: 1-47.

Zotin, A. I., 1972, *Thermodynamic Aspects of Developmental Biology,* S. Karger, Basel.

Section Two

Emerging Foci on Systems Research and Practice

THE HISTORY ON SOCIOCYBERNETICS, RC51, OF THE INTERNATIONAL SOCIOLOGICAL ASSOCIATION

Richard E. Lee

Fernand Braudel Center
SUNY-Binghamton
Binghamton, NY 13902-6000
USA

It is a distinct pleasure to have the honor of extending a welcome to this World Congress on behalf of the co-host organization I represent, the Research Committee on Sociocybernetics, RC51, of the International Sociological Association. "Socio" indicates a concern for the structure and development of human relationships and, indeed, I was trained as a sociologist. I do teach in a department of sociology and, in fact, my genealogy, by way of my own mentors, goes back through C. Wright Mills and, critically, the Columbia school of the 1950's. My association with that nomothetic discipline and its primarily present-oriented perspective, however, is mitigated by my long-time affiliation with the Fernand Braudel Center. This represents the collapse of the nomothetic-idiographic, and if you like, science-humanities or quantitative-qualitative, antinomy in both my personal disciplinary and institutional allegiances and my intellectual point of departure with which that collapse is bound up. As I was thinking about what to write, I was reminded that Ludwig von Bertalanffy, so many years ago, suggested that the systems approach he was advocating rendered moot this divide. So, from this deceptively simple premise, then, let me tell you something about the Research Committee on Sociocybernetics of the International Sociological Association.

Sociocybernetics has attracted a broad range of scholars whose departmental affiliations represent the entire spectrum of the disciplines, and I mean by that, from the humanities and the social sciences through the sciences, mathematics and engineering. Furthermore, the many countries of origin of these RC51 members attest to the wide international appeal of sociocybernetic approaches. Within this highly diverse community, there is wide agreement on some very general issues, for instance, on developing strategies for the study of human reality that avoid reification, are cognizant of the pitfalls of reductionism and dualism, and

Understanding Complexity, Edited by Ragsdell and Wilby,
Kluwer Academic/Plenum Publishers, New York, 2001

generally eschew linear or homeostatic models. Not surprisingly, however, there are also wide divergences in subject matter, theoretical frameworks and methodological practices.

The institutional history of the organization that eventually became ISA RC51 began in 1980 with the founding of an Ad Hoc Group within the International Sociological Association and proceeded with the organization of sessions at succeeding quadrennial World Congresses of Sociology: first in 1982 in Mexico City, then in 1986 in New Delhi, in 1990 in Madrid, and in 1994 in Bielefeld. RC51 became a Thematic Group and then a Working Group at the New Delhi World Congress and in early 1995 it elected its first ever board. The statutes for the group were written in 1995 and the publication of a biannual Newsletter was initiated in January 1996. In recognition of its extraordinary success--membership grew from some 30 members in early 1995 to 240 in 1998--the group was promoted to the status of Research Committee in less than the normal four-year period at the 1998 World Congress of Sociology in Montreal.

Since June 1998 RC51 has had its own website at the University of Zaragoza (http://www.unizar.es/sociocybernetics) where a detailed description of the group's activities, past, present and future, may be found. The website contains: abstracts of papers presented at the 1998 World Congress of Sociology in Montreal, of papers presented at the 1999 annual meeting in Kolimbari, Greece, and of papers presented at the 2nd International Conference on Sociocybernetics in Panticosa, Spain. Moreover, there is a 400-item bibliography on sociocybernetics, along with personal website addresses of some 65 RC51 members, detailed news about upcoming conferences and other activities, and links to websites in related fields.

I would like to invite everyone to visit us at the website, especially as we will shortly begin posting details pertaining to the organization of next year's 3rd International Conference on Sociocybernetics to be held in Léon, Mexico and the 2002 World Congress to be held in Brisbane, Australia.

RC51 also publishes the electronic *Journal of Sociocybernetics.* Soon to be available for downloading from the website twice yearly in the Spring and the Fall, it has now become the official publication of RC51 and invites submissions from both members and non-members.

The intellectual roots of sociocybernetics reach back to the rise of a whole panoply of new approaches to scientific inquiry beginning in earnest in the 1940's. These included General System Theory, cybernetics and information theory, game theory and automata, net, set, graph and compartment theories, and decision and queuing theory as strategies in one way or another appropriate to the study of organized complexity. Although today the Research Committee casts a wide net in terms of appropriate subject matters, pertinent theoretical frameworks and applicable methodologies, the range of approaches deployed by scholars associated with RC51 reflect the maturation of these developments. Here we find, again, GST and first- and second-order cybernetics; in addition, there is widespread sensitivity to the issues raised by "complexity studies," especially in work conceptualizing systems as self-organizing, autocatalytic or autopoietic. "System theory" (in the form given it by Niklas Luhmann) and world-systems analysis are also prominently represented within the ranks of RC51.

Although some would argue that models developed for the study of complexity can be usefully appropriated for the study of human reality (and the body of theoretical and empirical work in this area is growing), it seems to me that there is a larger issue here. On the one hand, the emphasis in complexity studies on contingency, context-dependency, multiple, overlapping temporal and spatial frameworks, and deterministic but unpredictable systems displaying an arrow-of-time suggests that the dividing line between the sciences and the social sciences is fuzzier than many might like to think. On the other hand, in the humanities, the uniquely modern concepts of original object and autonomous human creator have come under serious attack. The coincidence of these two phenomena substantiate our impression that across the disciplines there may be observed a new concern for spatial-temporal wholes constituted at once of relational structures and the phenomenological time of their reproduction and change.

It is, then, in this context of rich history and exciting possibilities that the Research Committee on Sociocybernetics of the International Sociological Association extends an open invitation to the totality of that ecumenical community engaged in the common quest to explain and understand social reality holistically and self-reflexively without forsaking a concern human values--human values not construed simply as a matter of individual ethics, but conceived as an integral part of a historical social science for our time.

SYSTEMS PRACTICE AT THE UNITED KINGDOM'S OPEN UNIVERSITY

Ray L. Ison

Systems Discipline
Centre for Complexity and Change
Technology Faculty,
The Open University,
Milton Keynes, MK7 6AA, UK

THE OPEN UNIVERSITY

The Open University (OU), the UK's largest University was established in 1969. It has since inception pioneered and supported open distance learning (i.e. all students have a course related tutor and entry is open to all at undergraduate level without prior qualifications). In 1998/99 the OU had 166,000 students of whom 28,000 were in taught post-graduate courses. In all the OU had 200,000 customers in 1998/9; Daniel (1996) has described it as a mega-university. Some other features of the OU include:

- An increasing number of students are from outside the UK (19% of students in 1998/99);
- The median age of new undergraduates is 34;
- Over 80,000 students study interactively on-line as some component of their courses;
- The Open University of the US (Inc) was established in 1998/99 as an independent subsidiary of the OU(UK);
- It is at the forefront of innovation with technology to support learning; £12.5 million has been allocated in 2000 for further expansion of e-university capacity;
- It is ranked amongst the top Universities in the UK for teaching quality (Technology, including Systems, received 24/24 in sector-wide teaching quality assessments in 1998);
- 26 subject areas have been recognised in UK-wide assessments as having research of international or national excellence.

Understanding Complexity, Edited by Ragsdell and Wilby,
Kluwer Academic/Plenum Publishers, New York, 2001

This paper relates the emerging focus on systems practice within the Systems Discipline at the Open University. Recent innovations have occurred in the changing environment of the University (Lane 1999), the higher education sector (Ison 1999) and the systems community more generally (Maiteny and Ison 2000).

SYSTEMS PRACTICE AT THE OPEN UNIVERSITY

The OU has been involved in 'applied' Systems teaching and research for 30 years - some 25,000 students have studied systems courses in that time (see Maiteny and Ison 2000 for an overview). The inspiration for Systems teaching came from the founding Dean of Technology who said:

"I felt that a concern for systematic study of the social and environmental aspects of technology was essential. Certainly environmental problems were approachable only by means of systemic and interdisciplinary methods and I felt convinced that any Faculty of Technology that did not concern itself with such problems could not claim to be either modern or responsible, whether socially or academically".

Systems and Design were therefore seen as key process-disciplines that were to work in close collaboration with the more conventional and analytical subject-disciplines of technology (e.g. electronics, mechanical engineering, materials engineering). The aspiration was that Systems and Design 'would contextualise and synthesise the subject-disciplines and act as catalysts for interdisciplinarity' (Holister, 1974). To a large extent this aspiration has been realised in one of two ways:

1. introducing systems thinking 'by stealth'. This has occurred through the contribution of Systems staff to the design and management of the Faculty's long-running Foundation, or level one course, 'Living with Technology' (OU code T102 and its predecessors T100 and T101) which 31800 students successfully studied from 1989 to 1999. Since its inception this course has situated technology in its social and systemic context. Presentation of this course ceased in 1999 but the strategy has been perpetuated in at least two of its replacement courses. For example in 2000, 12,000 students are studying the web-delivered course 'You, your computer and the net' (T171) which introduces some systems thinking concepts. A further 1200 students are studying T172, 'Working with our Environment: Technology for a Sustainable Future' which also introduces systems concepts.
2. developing and presenting overtly Systems courses since 1973 (Maiteny and Ison 2000). For example:
   - 10,795 students successfully completed the course 'Complexity Management and Change. A Systems Approach' (T301) from 1983 - 99; this course developed students ability to practise three systems methods: the failures method, the hard systems method and soft systems methodology;
   - 1,006 students had claimed the Undergraduate Diploma in Systems Practice award a year after its introduction in 1998 (this is gained by completion of 120 points of study of specified Systems courses, one-third of the points required for an undergraduate honours degree);
   - Over 300 post-graduate students have studied the core module: 'Environmental decision making. A systems approach' (T860) since the Environmental Decision Making (EDM) postgraduate programme was introduced in 1998.

Renewing the Curriculum and Ourselves

Over the last six years we have been engaged in a systemic action research process involving:

- curriculum innovation (see below);
- organisational and structural change

Following a process of review and sometimes intense internal negotiation, the Systems Department as it existed prior to 1998 was disbanded and replaced by the Centre for Complexity and Change (CCC) with three constituent Disciplines - Systems, Development Policy and Practice, and Technology and Manufacturing Management. As argued elsewhere (see Armson and Ison 1996) this constituted both a change of organisation and constituent structures and has facilitated the emergence of a new conversation about systems resulting in the formulation of a 'systems meta-project' comprising diverse activities (see Lane 1999).

- improving our own systems practice

The Open Systems Group is the umbrella under which we conduct our own systems practice, offering consultancy services, conducting our own staff development and managing these processes so as to develop and enhance our community of practice as if it were a learning organisation.

- building better relationships with our alumni

The Open University Systems Society, or OUSys, the alumni association of the OU had over 350 members as of late 2000.

- recruitment and appointment of new staff (from autumn 2000);
- reviewing the sustainability of our working practices (from 2000).

Systems Practice - The OU Pedagogical Model

Over the 30 years of our practice as systems educators a discernible pedagogical model has evolved. Four explicit strategies are recognisable:

1. Academics learned quite early that systems concepts need to be grounded as much as possible in the student's own experience. For this reason both continuous and examination assessment asks students to relate the systems thinking and practice in the courses to their own professional and personal contexts. We are aided in this strategy by the fact that most students are working whilst they study and they have a sufficiently rich life experience for the ideas to become meaningful.
2. Case studies of failure (e.g. IT innovations; the UK Child Support Agency etc) have proven to be a way of engaging students' involvement beyond their own experience. This was a lesson which was learned very early (e.g. Bignell et al 1977; Peters 1979) and which continues to be employed.
3. Diagramming (and other modelling) skills are developed and used as a means for students to engage with perceived complexity.
4. Other systems concepts, tools, methods, and methodological approaches are taught so as to develop skills in ***'formulating systems of interest.....for purposeful action'***. It is worth noting that in recent courses we have increasingly emphasised that purposeful action has both rational and emotional elements.

In the third level course presented for the first time in 2000 (Managing Complexity. A Systems Approach - T306) the metaphor of the systems practitioner as juggler is introduced to explicate ideas about systems practice (Figure 1).

Our courses are designed to develop students' abilities to practise systems. For example in T306, the idea of practice, or practising, is initially a very general one. The dictionary definition of practise is to carry out or perform habitually or constantly . . . to carry out an action. It is assumed that students have some role in which they practise and that most people occupy a number of roles, in their work or in their community. The Course Team assume that this is a role in which a number of issues that need dealing with, improving, resolving, or obviating are experienced. Figure 1 provides the basis for unfolding the metaphor of the juggler who represents the systems practitioner. The case is made that an effective practitioner has to continuously think about, and act to maintain, four elements (the four balls in the air); the processes of Being a practitioner, the situation being Engaged with, putting the approach taken into Context and Managing in the situation.

Figure 1. The Systems Practitioner (P) as juggler, juggling the processes of BEING a systems practitioner: the B ball; the situation the practitioner is ENGAGING with: the E-ball; the systems practitioner CONTEXTUALISING the approach they are taking: the C ball; and MANAGING in the situation: the M ball. (Source: Open University 2000, by permission).

Systems Practice - Our Domains of Focus

One of the outcomes of our review over the period 1995-98 was to consolidate our systems practice on three particular domains. This has subsequently affected our course design and our staff recruitment strategies. The three domains are:

(i) Developing Information systems
(ii) Environment and Sustainable Development
(iii) Managing in Organisations

Three themes can also be seen to cut across these domains, though it is unlikely that a consensus exists on this interpretation:

(i) critically reflective systems practice

(ii) the design of learning systems

(iii) considering and managing technology within its social and environmental context

Our recent courses and programmes as well as planned new programmes exemplify this new focus.

Our Main Systems Courses from 2001

Lane (1999) reports a heuristic model around which our new curriculum has been developed. Our strategy has been to develop two new core Systems courses at Levels 2 and 3 which when successfully completed can be used to claim an Undergraduate Diploma in Systems Practice. Rather than developing a stand-alone undergraduate named degree in Systems Practice we have attempted to situate Systems courses in as many named degrees as possible. For example the Systems courses will be part of a new named degree in Business Studies and negotiations continue for entry of the courses into a named degree in Environmental Studies. Another possibility is in the area of Health and Social Welfare.

The model reported by Lane envisages the curriculum progressing from a focus on Systems Thinking and enabling students to search for system in a number of 'messy' situations followed by a third level course focusing on systems practice using a range of systems methods for engaging with complexity so as to design and plan systemic change. A postgraduate programme in Systems Practice is planned which will enable students to explore the philosophical roots of systems thinking as well as further developing their systems practice.

Systems Thinking. Principles and Practice (T205). This second level course is a 60 point course (one year of study) which will commence in 2001. It consists of three sets of resource materials in conventional printed form that are drawn together in a teaching programme that is delivered via the Web and FirstClass). The use of the Web allows the inclusion of interactive and dynamic teaching techniques that are not possible in print, and are particularly appropriate for this subject area. It also allows the structuring and restructuring of the teaching much more responsively. The printed resources include:

- Training resources on systems thinking, presented in three packs: T551: Systems thinking and practice: a primer; T552: Systems thinking and practice: diagramming (which also includes video- and audio-cassettes); T553: Systems thinking and practice: modelling (which also includes a video). These packs represent a major innovation within the context of the OU as they are generic teaching materials designed for use in all of our courses as well as for sale.
- Concept files derived from various domains, but principally concerned with organisations entitled: Learning, thinking and doing; The individual (workplace and self-development); Groups and teams at work; Managing within organizations; Networks and the organizational environment.
- Case files concerned with: e-commerce and other social issues associated with supermarkets; managing household waste (an example of environmental decision making); the millennium bug or Y2K (as an example of risk management); homelessness. These also share a video-cassette.

The Web-based teaching programme is called:

- The T205 WebZone. It forms the teaching 'spine' of the course, and consists of six consecutive blocks of study activities, each ending in an assignment. They take students through several cycles of systems analytic activity, showing how to take a confusing and complex situation, develop a coherent, practical, view of it, and present that view to others.

Managing Complexity. A Systems Approach (T306). This 60 point third level course, first presented in 2000, applies recent and innovative developments in systems thinking to areas such as information systems, organizational change and learning, sustainable development and the environment and professional practice. The course develops skills to think differently and creatively about issues of perceived complexity, and enables students to appreciate and manage these issues in ways that lead to improvement. The content of the course is practical, chosen to enable students to use an action-learning cycle to intervene usefully in the world. The structure is holistic, exemplifying the same action-learning cycle:

- Finding out what's going on
- Thinking about what to do about it
- Doing what you've thought about
- Reflecting on the outcomes

The course draws on and extends a range of approaches to complexity management that have been developed by internationally recognized systems practitioners. They include the soft systems method, the viable systems model and the hard systems method. The introductory Block 1 introduces a case study that exemplifies organizational failure. Students are required to analyse it, drawing on their own experience and a range of systemic diagramming tools. These begin to equip students with some of the preliminary skills of formulating a system of interest and becoming a reflective systems practitioner (a perspective influenced by second-order cybernetic understandings). Block 2 shows how to create and review information systems in order to improve their relevance. Block 3 extends the systems-development methods of Block 2 to include the viable systems approach. Block 4 examines practical ways of managing sustainable development by taking a learning system approach and adapting mode 2 SSM for use in multiple stakeholder settings. Finally, Block 5 shows what can be gained by making the effort to reflect on the systemic outcomes of one's actions.

Experiencing Systems (TXR248). This is a 10 point for credit course based on our long-running summer school which until 2000 had been an integrated component of our courses. The residential component is the central element of the course. Before attending the school students are required to work through two of the three Systems Packs (T551 and T552 - see above). At the school students are taught the general structure of a systemic analysis, and are offered a range of scenarios within which they can experience the practical problems and benefits of adopting an holistic approach to the analysis of complex problems. After the school, students complete an assignment that tests their understanding of the methods and gives them an opportunity to reflect on what they have learnt. Paton (2000) outlines the course in more detail. We envisage this course being adapted to a range of contexts in the future including consultancy situations.

Environmental Decision Making PG programme. This programme has been established to meet the challenges of decision making in environmental and sustainable development contexts. The programme aims to meet the needs of specialists, managers and the public for postgraduate-level environmental education that can address the complexities of environmental issues. Everyday concerns of environmental protection, natural resource management and waste disposal, and rapidly changing environmental legislation and policy are considered in global and local contexts. Courses teach the skills individuals need to unpack the issues and participate creatively in the process of making environmental decisions - in all sectors of society. Students are equipped to use systems or holistic tools and techniques to deal with environmental issues relevant to their own lives and careers. The emphasis is on interdisciplinary interaction.

Students without a prior degree enter the programme by completing an Advanced Diploma in EDM; graduates initially complete 120 points of study including two core modules to gain a Postgraduate Diploma in EDM. To gain a MSc a 60 point Dissertation is then completed.

The Future..??

As with many people in early 21st century organisations we experience considerable stress in the workplace. In part this comes from a period of intense review and subsequent commitment to innovation. However there are longstanding issues, both local and institutional, which raise questions about the sustainability of our working, and thus family lives. For this reason we are actively embarking of an inquiry to examine how we might work more sustainably and ethically (by acting to increase choices for ourselves and our students).

Through such initiatives as the Open Systems group we are also attempting to enhance our own systems practice and to strive to narrow the gap between our espoused theory and our theories in use.

The dynamics of higher education and the increasing prospects of its globalisation mean that all academic groups have to scan the environment with a view to identifying actions that maintain the quality of their 'system-environment relationship'. In the future we will have to make trade-offs between innovations and the health of our working community. Potential innovations now being explored include a MSc in Systems Practice, a MSc in Information Systems (or some new variation of this). As well we have a desire to engage with others in building communities of:

- systems practice
- learning-resource developers for capacity building in systems practice (e.g. by following an open source model of innovation - see Naughton 1999 for a description).

CONCLUDING COMMENTS - SOME 'BIG' QUESTIONS AROUND WHICH AN INTERNATIONAL CONSORTIUM IN SYSTEMS PRACTICE MIGHT ORGANISE

As outlined by Maiteny and Ison (2000) the institutionalisation of systems thinking and practice within the academy is perhaps now more precarious than at any other time in the last fifty years. It is ironic that this should occur when the interest in, and seeming need for, systems practice in the general community has heightened. It would thus seem that there is a need to explore new institutional forms for capacity building for systems practice. Perhaps the Internet will facilitate the emergence of new communities of practice and thus new, co-operative, institutional forms. Should such consortia emerge I would suggest that there are at least five big questions that need to be brought into the conversation:

1. What constrains or enhances the translation of systems thinking into systems practice and new modes of systems thinking?
2. To what extent does the reification of current 'first-order' conceptions of knowledge, information and 'effective communication' constrain organizational change and the evolution of new communities of practice (for further explication of these ideas see Ison and Russell 2000)?
3. Is it possible to manage for 'self-organization' and emergence and if so how? (And what would 'to manage' mean in this context?)
4. What constitutes an effective pedagogy for building capacity in systems thinking and practice?
5. What constitutes ethical systems practice particularly in relation to the issues of global sustainable development and the threat of enclosure of the 'knowledge/information commons'?

In particular, questions relating to the design of contexts for systems thinking to develop and emerge into systems practice have for too long been ignored within the systems community. It is my aspiration that the Open University will play an active part in addressing these questions in co-operation with others and that together increased capacities to act systemically can be developed.

Acknowledgements

This chapter is dedicated to the memory of Graham Paton, a member of the Systems Discipline and the OU Technology Staff Tutor group who was tragically killed in a car accident a week prior to the ISSS conference. He was to have been with us in Toronto. The contribution of all staff in the Systems Discipline and EDM programme to the activities reported here is also acknowledged. They include: Rosalind Armson, Carolyn Baxter, Simon Bell, Chris Blackmore, Susan Carr, Ray Corrigan, Rita Daggett, Jacqueline Eisenstadt, Amber Eves, Jim Frederickson, Pam Furniss, John Martin, Dick Morris, Laurence Newman, John Naughton, Roger Spear, Tony Wright. My special thanks to Christine Blackmore for organising the abstract and to Pat Shah for developing the style sheet for the paper.

REFERENCES

Armson, R. and Ison, R. L., 1995, Systems and the 'polo mint' metaphor, in: *Critical Issues in Systems Theory and Practice*, K. Ellis, A. Gregory, B.R. Mears-Young, and G. Ragsdell, eds., Plenum, New York.

Bignell, V., Peters, G. and Pym, C., 1977, *Catastrophic Failures*, Open University Press, Milton Keynes.

Daniel, J.S., 1996, *Mega-Universities and Knowledge Media*, Kogan Page, London.

Holister, G. S., 1974, A view from the Technology Faculty, in: *The Open University Opens*, J. Tunstall, ed., Routledge and Kegan Paul, London, pp. 149-152.

Ison, R.L., 1999, Guest Editorial: Applying systems thinking to higher education, *Systems Research and Behavioural Science* 16:107-112.

Ison, R.L., and Russell, D.B., eds., 2000, *Agricultural Extension and Rural Development: Breaking Out of Traditions*, Cambridge University Press, Cambridge, UK.

Lane, A.B., 1999, Teaching Systems at the Open University: reconceptualising the curriculum by creating meaningful conversations, in: *Synergy Matters: Working with Systems in the 21st Century*, A Castell, A.J. Gregory, G. Hindle, M. James and G. Ragsdell, eds., Plenum, New York.

Maiteny, P.T., and Ison, R.L., 2000, Appreciating systems: critical reflections on the changing nature of systems as a discipline in a systems learning society, *Systems Practice and Action Research* 16:4

Naughton, J., 1999, *A Brief History of the Future: The Origins of the Internet*, Weidenfield and Nicolson, London

Open University, 2000, *Managing Complexity: A Systems Approach (T306) - Block 1*, Open University, Milton Keynes.

Paton, G., 2000, The process of experiencing systemic thinking: the evolution of a novel short residential course for the Open University, in: Proceedings of the 45th ISSS Conference, Toronto, Canada, Allen, J. and Wilby, J. (eds.), ISSS, Cookeville, TN.

Peters, G., 1979, On systems methodology, in: *Improving the Human Condition: Quality and Stability of Social Systems*, R. Ericson, ed., Springer, New York.

THE WOSC AND SOME VIEWS ABOUT CYBERNETICS

Robert Vallée

World Organisation of Systems
and Cybernetics
2, rue de Vouillé
75015 Paris, France

THE WOSC

The "World Organisation of Systems and Cybernetics" (in short WOSC) is a federation of associations or institutions concerned with systems or cybernetics, and, for the sake of requisite variety, it has two official languages: English and French. Son nom dans cette dernière langue est "Organisation Mondiale pour la Systémique et la Cybernétique". Its logo represents schematically the Watt's governor, which gives a classical example of negative feedback in engineering as well as a symbol of many kinds of regulations we can observe in living systems and in other fields of science.

The WOSC was founded in 1969 by Prof. J. Rose, who has been its Director-General up to 1987. The link between WOSC and cybernetics is emphazised by the fact that its President in memoriam is Nobert Wiener. Its President is Prof. Stafford Beer (UK), its Honorary Director-General is Prof. J. Rose (UK). The membership of WOSC is completed by a Director-General, Prof. Robert Vallée (France), a board of Directors, the affiliated associations or institutions and a council. Among the affiliated bodies let us quote: "Association Française de Science des Systèmes Cybernétiques, Cognitifs et Techniques" (or AFSCET), "Homeostatics of Living, Technical, Social and Ecological Systems" (Russia), "The Society of Management Science and Cybernetics" (India), "Cibernetica Scolastica" (Italy). Links have also been established with "Union Européenne de Systémique" (UES), which is a federation of European societies devoted to systems sciences.

The purpose of WOSC is to promote systems theory and cybernetics, mainly through exchange of informations between its members, the publication of a journal and the organisation of international congresses. The official journal of WOSC is "Kybernetes", founded in 1971 by Prof. J. Rose. Its Editor-in-Chief is now Prof. Brian H. Rudall, also

Understanding Complexity, Edited by Ragsdell and Wilby,
Kluwer Academic/Plenum Publishers, New York, 2001

Director at WOSC. It is published by MCB University Press (UK) and has 10 issues per year completed by special ones. It can be found on line via Internet. The "International Congress of Cybernetics and Systems" of WOSC has been held every three years since 1969. The first one took place in London, then it was in Oxford, Bucharest, Amsterdam, Mexico City, Paris, London, New York, New Delhi, Bucharest and Uxbridge in 1999. Moreover, WOSC has instituted an honorary Fellowship with 25 recipients and also a "Norbert Wiener Memorial Gold Medal", which has been presented to eight people. Dr. Candace Pert is the last to have received this distinction (during the evening organised by Prof. Stafford Beer in the name of WCSS). The other most recent recipients are Prof. Heinz von Foerster at the occasion of the annual meeting of the "American Cybernetics Society" held in Chicago in 1995. In 1990, for the 7th international congress of WOSC, in New York, the gold medal was presented to the recipient by Dr. Margaret Wiener Kennedy, daughter of Norbert Wiener.

The WOSC has also constituted a "Norbert Wiener Institute of Systems and Cybernetics" under the direction of Prof. Brian H. Rudall. Its aim is to help cooperation between associations or institutions affiliated to WOSC and to facilitate diffusion, for example, by publication in "Kybernetes" of research work on subjects related to systems or cybernetics such as, among others : viability, self-organisation, epistemology, bio and socio-cybernetics. Of course, WOSC has a website, which is managed by Dr. Alex Andrew, also Director at WOSC for international affairs.

ABOUT CYBERNETICS

It is well known that Norbert Wiener (1894 - 1964) and Waren McCulloch (1898 - 1969) had different views about cybernetics and that if Leibniz was for Wiener the patron saint of cybernetics, for McCulloch it was Descartes (McCulloch, 1974, posthumous publication). In fact, these positions are more complementary than opposite.

What attracted Wiener to Leibniz was his attempt of a kind of calculus of logics, or "calculus ratiocinator", once foreseen by Ramon Llul under the name of "ars magna" : "If I were to choose a patron saint for cybernetics out of the history of science, I should have to choose Leibniz" (Wiener, 1948). This calculus of logics, which had to do with binary numeration, was in fact never achieved. It was akin to the conception by Leibniz of a calculating machine that performs better than that built earlier by Pascal. Choosing Leibniz as patron saint of cybernetics was, from the part of Wiener, to put emphasis on an important, but not exclusive, aspect of cybernetics, that of the computer seen as the germ of a reasoning machine and a source of inspiration despite dangers of misinterpretations.

The choice of McCulloch in favour of Descartes may look surprising if we remember Descartes' bad reputation if not in cybernetics but in systems theory. A reputation which is partly due to misunderstandings of his thought, particularly about decomposition into parts which concerns, in fact, logical difficulties and not physical objects. Anyway, McCulloch and also Heinz von Foerster (Foerster, 1973) observed that, in his treaty on man, Descartes had a premonition of what is known now as reflex arc and negative feedback. Using a modern terminology to express Descartes' ideas, we consider a man approaching his foot close to a fire. A first signal is transmitted from the foot and, consequently, another signal comes back and acts upon the leg, making the foot escape the fire, maintaining, even if the fire and the foot move, an acceptable perception of heat and so a reasonable distance between them. So

McCulloch concluded : "As far as I know, this is the first use of the notion of inverse feedback, and so of the reflex" (McCulloch, 1974, posthumous publication). Another cybernetical aspect of Descartes' conceptions, in the same book on man, has to do with what we call encoding. Descartes says that what is transmitted, we would say information, is independent of the form under which it is transmitted, for example in the case of vision. McCulloch sees there what he calls "the first coding theorem" (Mc Culloch, 1974). We could add that this point of view of Descartes is not far from the idea of Ferdinand de Saussure according to which the thing signified may be represented by an arbitrary significant.

The interest of Leibniz and Descartes in very different aspects of what belongs now to cybernetics is representative of tendences of the human mind, which is attracted sometimes by algebraic considerations (here a possible logical calculus), and at other moments by topological views (here network connexions). Seen in another way, we have, on one side, the local aspect of algebra and, on the other one, the rather global view of topology. Using the words of Bourbaki, we can say that the human mind is tempted by the "demon of algebra" and also seduced by the "angel of topology" or attracted both by the discrete and the continuous.

These two fundamental categories give a classification of cybernetical trends : the algorithmic one, connected to informatics and discrete modelling, and what we could call the analytical one, in the sense of mathematical analysis, proposing continuous models. All the binary concepts and Boolean algebra belong to the first category. The Nyquist criterion based on the theory of analytical functions of a complex variable, the notion of homeostasis as well as the concept of fuzzy sets which dissolves too perfect dichotomies, have to do with the second. All the research involving the so called formal neurons belongs to the first class, global views concerning the unity of a cybernetical system due to information transfers, the conception of the world network as a kind of being of gigantic dimensions (Vallée, 1952), as well as the idea of "requisite variety", belong to the second.

But there is a kind of paradox. When we want to stick to discrete considerations, we are soon confronted with continuous ones. For example, the so called Z-transform (acquainted to Fourier transform) of a discrete signal is an analytical function and so is perfectly continuous. On the contrary, when we prefer continuous concepts, such as a compact Hermitian operator on an Hilbert space, the set of its eigen-values may be discrete. In fact, "nature" does not care about our categories, which just represent our preferences or the natural slopes of our mind. Some people are algebraically minded, some other topologically. At an elementary level, some like to do calculations, other prefer geometry.

I must confess that my preferences go to the second aspect of cybernetics. It seems to me that an interesting purpose of cybernetics, in that direction, is to provide "epistemo-praxiological" models of how much a conscious being is able to know and modify both itself and the rest of the world (Vallée, 1955).

REFERENCES

von Foerster, H., 1973, On constructing a reality, in : *Environment Design Research*, 2, F. E. Preiser, ed., Hutchinson and Roess, Dowen.

McCulloch, W. S., 1974, Recollection of the many sources of cybernetics, ASC *Forum*, 6 :2.

Vallée, R., 1952, Cybernetics and the future of man, *Impact of the Science on Society*, 3 :3.

Vallée, R., 1955, *Cognition et Système, Essai d'Épistémo-Praxéologie*, L'Inter-disciplinaire, Limonest.

Wiener, N., 1948, *Cybernetics or Control and Communication in the Animal and the Machine*, Hermann et Cie, Paris, The Technology Press, Cambridge, John Wiley and Sons, New York.

PRESENTATION OF WIENER GOLD MEDAL TO DR. CANDACE PERT

Stafford Beer

Viable Technology
34 Palmerston Square
Toronto, Ontario, M6G 2S7
Canada

INTRODUCTION

The World Organization of Systems and Cybernetics, meeting in the context of the World Congress of the Systems Sciences held in Toronto July 16-22, had the great pleasure of welcoming Dr. Candace Pert on Tuesday 18th July to receive the Wiener Gold Medal. The President, Professor Stafford Beer, gave the following address.

ADDRESS

I am so grateful to see such a large audience here this evening, given that the programme of this conference offers eight alternative meetings beginning at this very moment. Some years ago, in an exactly similar situation, I was hurrying to my chosen venue when I noticed an eminent statistician sitting alone on a chair looking disconsolate. I paused to ask him whether there was not a single meeting out of eight worth attending. He said: 'That's not the problem. The problem is that I shall perforce miss seven excellent presentations, and in terms of standard deviations it seems to me that there is no significant difference between missing seven and missing eight. So I'm staying here'. Congratulations to you all on overcoming any such blockage, and a hearty welcome from everyone to Candace Pert.

Some twenty-five years ago Candace discovered the opiate receptor. This is a site in the cell that can recognize an opiate, typically a protein molecule, which is then anchored in the outer cell membrane to bind with substances such as neurotransmitters. There was confusion and disagreement at the time, as to whether the biochemical components even existed in the body naturally to create such outcomes.

The search to find the opiate receptor was one of dogged endurance reminiscent of the search for radium. Other scientists were searching too, but it was she who discovered a pair

Understanding Complexity. Edited by Ragsdell and Wilby.
Kluwer Academic/Plenum Publishers, New York, 2001

of amino acids constituting the critical peptide. This in itself was a discovery of major importance -- significant people in the field expected Candace Pert to be awarded the Nobel Prize. The non-story of that, and of how the hardly less significant Lasker Award for medical research [often endorsed by a later Nobel Prize] was awarded to three *men* -- men heavily underlined -- will appall but not surprise egalitarian scientists, especially if they remember the shocking events surrounding Rosalind Franklin and the discovery of the DNA molecule. You may read about this in Candace's book *Molecules of Emotion*, and very entertaining and exciting you will find it.

However, I do not expect a cybernetic conference to celebrate this discovery with particular enthusiasm. Please bear with me as I follow the peptide story a little further. All sorts of peptides were shortly discovered, and a whole new era was to begin. I suggest that we focus what was to happen on Candace Pert's own comments. Where about in the body would you expect to find opiate receptors? Obviously you would look in the brain itself -- the hypothalamus perhaps. Alternatively you would look in the limbic system. But when she looked comprehensively for 'her' peptides, she found them all over the place in the body. Think of finding concentrations of such peptides in the colon, as she did... so *that's* where "gut feelings" come from!

We move to the early '80s. The neuropeptides, it had reasonably been assumed, communicated across synapses in the nervous system. The assumption proved untenable.

Many of the neuronal receptors were inches away from the neuropeptides: how were they communicating, if not across the synaptic gap? A co-worker called Miles Herkenham found that, counter to the assumption of people working in the neurosciences, less than *two percent* of neuronal communication actually occurs at the synapse. This seems so absurd that for several years the result was ignored, and put down to errors of one sort or another. But Miles Herkenham was right all the time. He reckoned that the connection did not reside in the synapse-brain cells, but was determined by the specificity of the receptors. Candace Pert wrote: "the way in which peptides circulate through the body, finding their target receptors in regions far more distant than had ever previously been thought possible, made the brain communication system resemble the endocrine system, whose hormones can travel the length and breadth of our bodies. The brain is like a bag of hormones!"

At about this time, Francis Schmitt, who had originated the neuroscience research program at MIT, introduced the terminology of "information substances" to describe 'a variety of transmitters, hormones, factors, and protein ligands' -- ligands are various small molecules that specifically bind to a cellular receptor, such as the opiate receptor, thereby transmitting an informational message to the cell. This was exactly the concept that Candace Pert needed to advance her own work, and she embraced it enthusiastically.

Now there are three classically separated areas of medical biology:

- Neuroscience -- dealing with the brain and central nervous system,
- Endocrinology -- dealing with the glands,
- Immunology -- dealing with the spleen, bone marrow, and lymph nodes.

Each of these has developed independently of the others, and offers a classic case of the Hardening of the Categories to which my own writings complaining about reductionism in the systems sciences constantly refer. Explicitly, throughout the '50s, my attempts to incorporate the endocrine system into my neurophysiological brain model foundered on the rocks of disparate and incommensurable disciplines. So you will understand the excitement with which I received the discoveries that *Molecules of Emotion* unfolded. Instead of those three sciences, demarcated by their library shelves and dedicated journals, and instead of obediently following them into separate laboratories, we are presented with a unified system. It consists of a multi-directional network of communication, linked by informational carriers *at the molecular level.* It is surely delightful to contemplate the continuous molecular busyness that achieves wonders of intricate homeostasis -- while quite indifferent to the pompous definitions of academe.

May I urge you to read the book. I do not have the effrontery to dissertate at greater length, thereby spoiling Candace's own account and a thoroughly good read. But I must share with you a final insight that I met in the text with a squeal of joy. Candace explains that the unconscious mind of Freud is nothing other than the body itself -- operating at the molecular level to integrate every aspect of the self. In fact, her discoveries are pointing to the kind of holistic emphasis on the unity of being that is familiar in eastern philosophy. I see her helping to cross that East-West divide -- and that other chasm existing between science and philosophy.

Surely these are matters for high celebration. By the end of her book she is openly hypothesizing about connections not only between body and mind, but between body, mind, soul and spirit. Predictably, she will have a rough ride as do all holists in a reductive world. I should like to wish her well in those endeavours, and that she continue with the same brave-heartedness with which she confronted so much prejudice in the past. Meanwhile, her scientific demonstration of the molecular reality of informational substances -- the neuropeptides -- in continual interaction between body and mind is the great ***cybernetic*** triumph acknowledged tonight.

SUMMARY

I ascertained before I started that hardly anyone in this audience knew of Candace Pert, still less of the cybernetic triumph. How can this be? Surely it is because each of us here is pursuing the next step in the agenda s/he has elaborated within the confines of the paradigms that are already understood. This comment is not meant offensively. The research we are all doing, the development of the thinking we so far understand, are all worthy pursuits -- the backbone of scientific advance indeed. But as system scientists, are we constantly in search of systemic invariance? We should be. Do we ever consider taking time off to review systemic advances in unfamiliar fields? We ought to. If not, we are tacitly accepting the established paradigms of systems, tacitly resisting change and the hope of new visions. I doubt if anyone present actively wants to appear in that role.

I now call upon the Director General of the World Organization for Systems and Cybernetics, Professor Robert Vallee of Paris, to read out the citation.

The Executive Council of the World Organization Of Systems and Cybernetics has awarded the Norbert Wiener Memorial Golden Medal to

Dr. Candace B. Pert

in recognition of the relevance to cybernetic science
of her seminal work in molecular biology
considered as informational substances,
and of providing pioneering scientific insights
into an holistic view of the body-mind unity.

Signed by the President on Tuesday Eighteenth July 2000
[signed] Stafford Beer.

The President then presented the medal to Dr. Pert, who responded with a brief acknowledgment of thanks. However, she graciously agreed to join in an open discussion, and an enthralled audience remained in their seats until the President closed the proceedings an hour later than the advertised closing time.

REFERENCE

Pert, C., 1998, *Molecules of Emotion (Why You Feel the Way You Feel)*, Simon and Schuster, New York.

Section Three

Human Systems in the 21st Century

ON MISDIRECTED SYSTEMS

Russell L. Ackoff and John Pourdehnad

Institute for Interactive Management
Six S. Bryn Mawr Avenue, Suite 200
Bryn Mawr, PA 19010-3215

INTRODUCTION

Every system has one or more functions in one or more larger systems of which it is a part. For example, an automobile is part of a transportation system and its function is to carry people from one place to another. When Newton asserted that the universe, which contains everything, was a mechanical system, he was faced with an apparent paradox: How could everything be a part of something larger? This was not a problem for the genius of Newton. For him the universe was a machine that had been created by God to do His work.

The function that many systems preach is not the one they actually practice. This may be due either to an error or deliberate obfuscation. In either case the system winds up doing the wrong thing. Recall Peter Drucker's observation of the difference between doing things right and doing the right thing. This distinction is fundamental. The righter we do the wrong thing, the wronger we become. If we make an error doing the wrong thing and correct it, we become stronger. Not so when we pursue the right thing.

The examples and statistics we use in this paper are for the United States. However, we believe the conclusions we draw from them are valid for all well-developed countries.

ENERGY SYSTEMS

The "better" automobiles that are now being produced are clear examples of doing the wrong thing righter. They are dysfunctional in several different senses.

First, size: the early automobile was designed as a luxury item for the average American family that then consisted of 5.6 people. Today the automobile is no longer a luxury but a necessity for most people and the average family has dropped to 3.2 people. Furthermore,

Understanding Complexity, Edited by Ragsdell and Wilby,
Kluwer Academic/Plenum Publishers, New York, 2001

most households have two or more automobiles so it is reasonable to consider both a small urban automobile for use within cities and a larger one for use between cities. The average number of people in an automobile in an American urban area is 1.2. About 85 percent of the cars going from one place to another contain only one or two people.

If use of city streets were restricted to two-passenger automobiles that could go not faster than 40-50 miles per hour, the people carrying capacity would increase from 300-500%. Such vehicles would also reduce the second dsyfunctionality of automobiles, their production of congestion and accidents and the third, their contribution to air pollution. Bill Hampton (1999, p.164) writes:

> "According to the U.S. Energy Information Administration, the United States generates more than five billion tons of energy-related carbon dioxide annually, making it the world's largest single emitter of the gas. About one-third the nations' total output comes from the transportation sector, and its share has been growing faster than all others have since 1990.
>
> Blame it on America's urge to drive big, powerful vehicles that inherently consume more carbon dioxide than smaller vehicles do. "
>
> It is much better to do the right thing wrong than the wrong thing right.

HEALTHCARE SYSTEMS

A large number of our social systems and institutions are busily engaged in doing the wrong thing and trying to do it better, hence wronger. Consider the U.S. healthcare system. It has been deteriorating over the last few decades. Furthermore, the U.S. is the only well-developed country with less than universal healthcare coverage; about 40 million people are not covered. According to Joel R. Cooper (1999):

> "Currently more than 43 million Americans lack health insurance, and the number is expected to grow to more than 47 million within the next five years."

Despite the propaganda issued by our politicians the U.S. healthcare system is not the best in the world. Those who practice medicine realize this. Recently, American Medical Association President, Nancy Dickey (1998) told an audience that:

> "America's current healthcare system is a mess and needs to be made the top priority in Congress. Despite huge advances in medical technology and not withstanding efforts to take advantage of these breakthroughs without raising costs, no one is really happy with the current system. Patients, physicians, the people who pay for the care all are frustrated. We haven't got it right, our systems are out of balance it's a mess."

The World Health Organization ranks the United States well down the list. Holland which ranks first spends only about one-third as much per capita as the United States does:

"We spend an estimated 800 to 1 trillion dollars per year on healthcare. We have more doctors, more hospitals, and take more drugs than any other nation yet, the World Health Organization ranks us among the sickest. Health care expenditure growth in US was 3% per annum higher than the expenditure of all other goods and services. By the year 2030, if the rate of growth is not controlled, 30% of GDP will be absorbed by the health sector" (Inglehart, 1999).

Something is fundamentally wrong! What is wrong is that it is not a healthcare system, but a sickness- and disability-care system. Its servers are compensated for taking care of the sick and disabled. Therefore, their income depends entirely on the amount of sickness and disability available for treatment. Whatever the intentions of the individual servers, the system produces and preserves sickness and disability. There is ample evidence of this. Andrea Gerlin (1999 pp. A1 and A13):

"The chance of error occurring during treatment may be far greater than you had imagined. ... according to Lucian Leape of the Harvard School of Public Health, an estimated one million people are injured by errors during hospital treatment each year and 120,000 people die as a result of those injuries."

An excessive amount of testing is done in order to pay off the excessive number of MRIs and other expensive pieces of equipment. There is also evidence of unnecessary surgery and prescription of drugs. Lauren Neergard (1999) reports that:

"Surgical gaffes like amputating the wrong foot or deadly chemotherapy overdose make headlines. But patients may never hear of the more subtle errors, like a delay in diagnosis or testing that costs precious time to fight off disease. Medical mistakes are a stunningly huge problem, says a new report by the Institute of Medicine. It quoted studies estimating that at least 44,000 and perhaps as many as 98,000 hospitalized Americans die every year from errors. "

If the healthcare system were really to be dedicated to healthcare, primary-care physicians would be paid an annual fee to preserve one's health. They would then have to pay all medically-related expenses, for specialists, testing, drugs, and so on. Then the only way that physicians could make a living would be to keep their "patients" healthy. Of course steps would have to be taken to assure that patients receive the treatment they require, but this turns out to be relatively easy to do. (See Rovin et al., 1994).

CRIMINAL JUSTICE SYSTEMS

The criminal justice system provides another example of doing the wrong thing right. King (1999) writes:

"The U.S. prison population, 1.2 million not counting short-term jail inmates, is the largest in the Western world. A number of states are spending more on prisons than on schools.

In mid-year 1998, we in the U.S. had 1.8 million of our citizens behind the bars. That's up from 200,000 in the 70s and 74,400 in 1985. We grew by a factor of 2.4 in these 13 years. As early as 1996, there were already about 5.5 million adult residents of the U.S. in prison, jail, parole, or probation. We now imprison at least six to 10 times as many, per 100,000 population as most civilized countries. While we incarcerate 668 person per 100,000, England, France, Germany, Switzerland, Netherlands, Norway and Japan all incarcerate only between 37 and 100 persons per 100,000."

The system doesn't work. As Alistair Mant (1997) concluded:

> "If you have more of your citizens in jail or on remand than you have in college (as in the United States) then the big system cannot be functioning well."

The assumption on which the U.S. system operates is that the apprehension and incarceration of criminals reduces crime. Not so! A criminal who has been in prison and is released has a greater probability of committing another crime than he/she had when he/she went into prison, and it is likely to be one that is much more serious. Therefore, the more police the U.S. hires and the larger the prisons it builds, the more crime and more serious the crimes, committed.

> "Altogether, our annual layout for corrections is more than $35 billion, curving steadily upward even as crime rates drop. We are developing a powerful prison-industrial complex. A national growth industry exploiting today's hostility towards wrongdoers." (King, 1999)

To increase the percentage of criminals caught and imprisoned, in the current prison system, is to do the wrong thing righter. Prison is currently an institution for rendering punishment, getting revenge for infractions of the law - not rehabilitation of criminals, which it is supposed to be.

> "A New York State Unified Court System Committee, including seven distinguished judges, reported: It is apparent that each year thousands of persons who present no, or minimal threat, to public safety are being committed to state prison. And this is being done despite the exorbitant costs involved and despite the fact that community-based programs may well be as effective, or appreciably more effective in rehabilitating these offenders." (CURENY, 1999).

EDUCATIONAL SYSTEMS

Now consider another institution: the educational system, which is supposed to have learning by students as its objective. It clearly has a more dominant objective, teaching by faculty members. The incorrect assumption on which this is based is that being taught is an effective way to learn a subject. Unfortunately, this is not true for most subjects for most students: for them teaching is a major obstruction to learning. For example, consider language. We learn our first language without having it taught to us, but never learn a second language in school well enough even to order food in a restaurant.

On the other hand, it is apparent to anyone who has ever taught a class that it is the teacher who learns the most. Therefore, schools are upside down. Students should be teaching because this is a good way to learn, and so-called teachers should act as resources available to students to use as they see fir in their efforts to learn. Teachers have demonstrated a much greater ability to learn than to teach.

Lee Smalley (1989), in the editorial section to the *Journal of Technology Education*, writes:

> "Teachers generally teach us as they were taught, as well as continuing those techniques that seem to be consistent with peer approval and support from administration and parents. Therefore, the teacher as teller continues far longer than is appropriate in this communication age. We have successfully rejected the radio, telephone, and television as an instructional device and are doing the same with the computer ... teachers still stand and talk while students sit and listen. Teachers work too hard during class time and students are too passive ... teachers should let the technical devices do the easy work - transferring information - and leave the more difficult task of developing generalizations and providing a social context to the information learned by the teacher-student ... teachers should be helping students discover their talents and designing activities to encourage their strengths, rather than running an animal school where all the animals have to fly, climb, dig, run, jump, and hide."

Tests have been conducted which show that second grade students learn more arithmetic by teaching it to a computer than by received conventional computer instruction. Architectural apprenticeships produced three great American architects: Richardson, Sullivan and Wright. Since architecture has been taught in universities, not one of the same stature has been produced.

In a survey once conducted among American statisticians to determine who, in their opinion, had contributed most to their field in the recent past. Four were selected. Not one had ever had a course in statistics. D. C. Seeler et al. (1994) assert:

> "The instructor must create a learning environment which increases students' involvement in, and responsibility for, the learning process ... the role of the student and faculty member changes and both must be ready to accept those changes ... students must take responsibility for their own professional development and increase their level of participation in the process. They must change from simply dealing with the use of isolated facts to subjects becoming aware of the relevancy of the information and the immediate application of real situations."

Now consider universities, the declared primary function of which is to educate students. Not so! One of us spent two years on the Senate of a college in his university as a representative of the college of which he was a part. To escape boredom he kept records of the topics discussed. Only once in two years was a student matter raised. The concern of the faculty was the faculty not students. Despite its avowed function, the function universities actually carry out is to provide their faculties with the quality of work life and standard of living they want. For example, on the issue of tenure, Jim Collins (1999, pp. 4-5) writes:

> "Around 1910 an emerging mechanism or practice grew up called academic tenure. The idea behind tenure was: you can't get fired for what you think. It was a very important practice to support the core value of freedom of inquiry. But now as I look at universities, I find them protecting the current version of tenure - as if tenure is what is sacred. Tenure is not sacred; freedom of inquiry is sacred. And tenure is the means, thus far, by which we have supported that core value. I see universities getting stuck, and saying, "Well, we can't touch tenure because... " thinking tenure is part of the value system. It is not. It is a practice to support the value system, and if universities are to evolve, they must be willing to evolve the tenure system while preserving the core value."

Teaching is a price faculty members must pay for the quality of work life they want. And like most prices, they try to minimize it. The higher the status of a faculty member - hence the more control he/she has - the less teaching he/she does. The higher the status of a university, the fewer the average teaching hours of its faculty.

ORGANIZATION SYSTEMS

Universities are not alone. Most corporations are much the same. A while back a distinguished management scientist checked the most important decisions made by the executive office of a major corporation against its published list of objectives. He found more than 90% of the decisions violated at least one of the objectives. He observed that there were only two possible explanations. Either the executives were stupid or the published corporate objectives were not the real ones. He believed the second explanation was more credible. As a result he asked: What objective(s) would make the largest number of the decisions made optimal or near optimal? He found one objective which did this for most of the decisions: to maximize the quality of work life and standard of living of those who made the decisions.

It is obvious to one who examines the corporate jets, chauffeured limousines, executive dining rooms, lavish offices and assistance, health clubs, county club memberships, and other perks, that maximizing shareholder value is not an objective. Providing shareholders with a satisfactory return is a requirement, not an objective. As Peter Drucker once pointed out: profit is to a corporation what oxygen is to human beings; necessary for its existence, not the reason for it.

Let's probe corporate misdirection a bit more deeply. From the point of view of society, of which a corporation is a part, the corporation has two functions: to produce and distribute wealth. The wealth it produces is essentially the difference between the amount of consumption it makes possible and its own consumption. Its distribution of wealth takes a number of forms, payment of interest on debts, dividend on investment, purchases of goods and services, but most important, compensation for work.

Productive employment, therefore, is the only means available to a society for simultaneously producing and distributing wealth, and unemployment is a major destabilizing political force in a free-market economy. This is why countries like England and Mexico nationalized a large number of industries after World War II: not to increase profit but to maintain employment. For the same reason, the U.S. Government supported Chrysler when it was on the brink of bankruptcy: to preserve employment.

Therefore, from society's point of view, downsizing is dysfunctional, and, as Wysochi (1995) has pointed out, it is even dysfunctional from most corporations' points of view. According to Jonathan Lurie (1998):

> "The term 'downsizing' was coined to describe the action of dismissing a large portion of firm's workforce in a very short period of time, particularly when the firm was highly profitable."

Carpenter (1996, pp. 36-39) noted:

> "Downsizing is often a short-term solution to increase profits and reflects how corporations focus on financial figures instead of taking a holistic view."

When we have pointed out these deficiencies to corporate executives, many are outraged. They ask: What do you expect us to do when we have extra people in an increasingly competitive economy? Good question. A good answer is: Create productive employment for them. It can and has been done.

Downsizing might be possibly excusable if it were successful. Ralph Estes (1997), a former CPA with a Big-6 accounting firm has gone to battle with corporations that are guilty of several sins:

> "... skimping on safeguards that would prevent worker injuries; emitting harmful pollutants that result in illness and birth defects; announcing mass layoffs despite record earnings and whopping increases to top management; and moving factories outside the country to reduce labor costs. Why are people in corporations doing these bad things? Because the only score that counts today is the bottom line - the gains and losses to stockholders."

These observations were echoed by Peter Drucker (1991, p. A10):

> "Scores of large organizations - businesses of all kinds but also government agencies, hospitals, and universities - have sharply cut staffs in these past few years. But few have realized the expected cost savings. In some cases costs have even gone up. In many more performance has suffered ... ".

Creative alternatives to downsizing are available. For example, and this is only one of many we could cite, one company that was on the verge of bankruptcy found it could purchase transportation of its products to its distributors from outside trucking companies at a lower cost than its own trucking unit could provide it for. Rather than fire drivers and others in the transportation department of more than 400 employees, the CEO called all the members of the department into a meeting and told them why it was necessary for the company to outsource its transportation. Then he showed them how much it would cost the company to discharge those present. He then suggested that the money involved could be pooled, and that together with a bank loan (which he had tentatively arranged) the employees could buy the department. They could then establish a new trucking company, to which he would give an extended contract for transporting the company's products on condition that it charge no more for doing so than commercial carriers. The new company was established and has lived happily ever after. No-one lost employment.

Another example cited in Ackoff and Pourdehnad (1997, p. 8) is about how Toyota Motors dealt with excess labor cost problems:

> "Toyota Motors, handicapped by an excess of employees, asked them how to reduce costs, raise productivity, and improve morale. In fact, through four years of recession, Toyota not only upheld the tradition of lifetime employment in Japan; it actually turned to its workforce of 70,000 for help."

Finally, Karnitschnig (1999, p. 72) reports that:

> "What started as a way to keep engineers busy in periods when demand for its sports cars sagged has become a pillar of Porsche's business and a key part of its growth strategy. The success of the business has also secured Porsche's status as Europe's last independent carmaker. Indeed Porsche has turned the rent-an-engineer business into a cash-cow."

CONCLUSION

To the extent that systems thinkers focus on efficiency (doing things right) without considering effectiveness, they are behaving irresponsibly, if not immorally. They have an obligation to make the organization with which they are working aware of its actual functions in contrast to the one or ones it avows. Revealing the truth about itself to an organization is a major contribution to its welfare that a systemalist can provide. Without such self-knowledge transformations that lead to 'thrival', in contrast to mere survival, are very unlikely. It is only when what is practiced and what is preached are consistent that the right thing can be done right.

REFERENCES

Ackoff, R. L. and Pourdehnad, J., 1997, The irresponsibility and ineffectiveness of downsizing, *Systems Practice*, 10 (1), 8.

CURE-NY, 1999, A Dialog on Criminal Justice, *Criminal Justice Reform*, http://www.bestweb.net/~cureny/dialog.htm

Carpenter, T. R., 1996, Corporate anorexia: a dangerous epidemic, *USA Today (Magazine)*, July, 36-39.

Collins, J., 1999, The paradox of continuity and change, *Drucker Foundation News*, 7 (3) 4-5.

Cooper, J. R., 1999, Physician groups say: all Americans must have health insurance, *The Medical Reporter*, June.
http://www.medicalreporter.health.org/tmr0699/universal-health.html

Dickey, N., 1998, *The New York Times*, December 17,
http://www.fortht.com/~chronicl/prop1.htm

Drucker, R., 1991, Permanent cost cutting, *The Wall Street Journal*, January 11.
Estes, R., 1996, *Tyranny of the Bottom Line: Why Corporations Make Good People Do Bad Things*, Berrett-Koehler, San Francisco.
http://www.trainingsupersite.com/wconnect/wc.dll?tss3~showabs~001127
Gerlin, A., 1999, Study slams medical error rate in nation, *The Philadelphia Inquirer*, November 30, A1 and A13.
Hampton, B., 1999, Environment, section on smart cars 2000, *Business Week*, November 15, 164.
Iglehart, J. K., 1999, The American healthcare system - expenditures, *New England Journal of Medicine*, January 7, 340 (1).
http://www.forthrt.com/~chronicle/prop1.htm
Karnitschnig, M., 1999, That van you're driving may be part Porsche, International Business Section, *Business Week*, December 27, 72.
King, R., 1999, It's time to open the doors of our prisons, *Newsweek*, April 19. http://www.newsweek.com/nw-srv/issoe/161999/printed/us/dpt/my/my02161.htm
Lurie, J. M., 1998, *Downsizing, A dissertation*, Princeton University, Department of Economics.
Mant, A., 1997, *Intellectual Leadership*, Allen and Unwin, St Leonards, NSW, Australia.
Neergaard, L., 1999, Report cites deadly medical errors, *The Associated Press*, November 30. http://wwwcpages.dynip.com/nsi/news%20for%20web/medical%20news.htm
Rovin, S., Jeharajah, N., Dundon, M. W., Bright, S., Wilson, D. H., Magidson, J. and Ackoff, R. L., 1994, *An Idealized Design of the US Healthcare System*, Interact, Bryn Mawr, PA.
Seeler, D. C., Turnwald, G. H. and Bull, K. S., 1994, From teaching to learning: Part III, lectures and approaches to learning, *Journal of Veterinary Medical Education*, 21 (1).
Smalley, L., 1989, Images of schools: 2020, possible, probable, preferable? Editorial Section, *Journal of Technology Education*, 1 (1), Fall.
Wysochi, B. Jr., 1995, Some companies cut costs too far, suffer corporate anorexia, *The Wall Street Journal*, July 5, p.1.F.

THE COMPLEXITY OF PERFORMANCE AND THE INSTRUMENTS OF MASTERY

Paul C. Wohlmuth

Institute for Law and Systems Research
University of San Diego School of Law
Alcala Park
San Diego, CA 92110 USA

INTRODUCTION

The range of our responses to situations is circumscribed by the influence of standards on human behavior. Depending on *how* they are shaped, expressions of standards have varying impacts on performances, from facilitation to oppression. I consider here the relationship between the embodiment of standards and a critical balance between automatic and creative elements in performance.

For experimental purposes in the kind of laboratory I will undertake to identify, I will try to evoke here a palpable sense of the pressure on performance produced by the range of ways in which standards can be embodied. In doing this, I will call attention to the variety of means enlisted for the expression of standards, the time scales in which each operates, and how they are integrated. I will posit a continuum of resulting configurations, at one end of which I will place what I call "instruments of mastery" (IOM's) and at the other what I call "dogma." I will show how the character of the blend between automation and creativity in formulations of performance is illuminated by locating relevant expressions of standards along this continuum.

As illustrative of IOM's, I have selected the sonata form in music and the automobile driver's handbook. I might have selected any of a large number of such instruments, more or less familiar to the reader. An outline, memorandum or other document, in hard copy or computer-accessible, a figure, a voice recording, a device, or a procedure or mantra one can silently rehearse could have served as well. Consultation with these instruments can continue to be useful after the impact of initial exposure. Indeed their availability for revisiting leads to a sense of mastery that I will be attempting to evoke. These renewals may

Understanding Complexity, Edited by Ragsdell and Wilby,
Kluwer Academic/Plenum Publishers, New York, 2001

function either as repeats, extensions, or elaborations of the original exposure. A periodically updated version may be particularly useful and reassuring, so long as the basis of earlier mastery is securely carried forward.

The most striking yet unsung aspect of IOM's is how they serve, often repeatedly, as benchmarks for engagement in the formulation of competent and creative performances. Once introduced, they tend to remain unobtrusively in play, as their contribution to mastery requires internalization. We are served by some but not a lot of awareness of their presence. Their character is syntactic, not unlike the basic vocabulary and grammar we effortlessly carry forward from childhood. While their content is and must be quite stable, they are subject to pressures for reformulation. Occasionally, cumulative pressure of this kind will tend toward major reworking of the instrument, though this is more likely to occur gradually and even imperceptibly, rather than dramatically. Quite ephemeral instruments, like initial notes or a sketch for a paper, play, painting or musical composition share these characteristics, although special skill may be required to distinguish any such instrument from the performance that it is serving to facilitate, as with scores, lyrics and scripts. But the *source* of pressure is always performance.

PERSONALIZATION

The quality of our relationships to IOM's is crucial to the quality of performances. Their origins are likely to be prototypes established in institutional and public domains. But it is crucial to the utility of IOM's that they be *personalized.* I define personalization as the complex process in which an actor balances a variety of internal and external influences in shaping his/her performance. While development and consummation of IOM's tends to be relatively unthoughtful, more assured mastery requires periodic attention to them. It is their personal adaptability that functionally distinguishes them from the less user-friendly instruments that they tend to come into relation to, of which I will say more shortly.

IOM's having long lines of traceable descent have been forged out of relentless personalization, often in the absence of any corresponding awareness or intention. Sonata form, extant for centuries to help shape musical compositions, illustrates. Emergence and continual transformation of the symphony, concerto and other compositional types has always stood on the foundation of the sonata (and related prototypes) despite crucial innovations built upon it by well-known and obscure composers. As with all other IOM's, there is nothing inevitable about sonata form beyond the fruition of something like it; yet once put in place, amidst extended trial and error, it has proved friendly and enduring. It has managed to help shape the creative disposition of composers and musicians by meshing automation and creativity in a manner facilitative of the more personalized expression essential to the realization of musical genius (Newman, 1972a, 1972b, 1969). The success of this and other formative foundations for artistic creativity underscores simultaneously both the central role of IOM's to craftsmanship and the delicacy of the balance to be struck in the process of personalization.

I select sonata form not only as one of the more enduring IOM's, but also because it provokes the extraordinarily difficult issue of the bearing of pedigree on the authority of standards in the formulation of performances. It is an unresolved issue in cognitive and other behavioral sciences how minimal a repository of standards can be while serving as foundational to performance rendering, and how this quantum of structure is built. What seems clear is a tendency, probably set loose by the very centrality of IOM's to human creativity, for standards to crystallize *beyond the point* of maximum utility to gratifying performances. Apparently because the subtlety of their operation makes the authority of IOM's unstable, these more elaborated institutionalized standards gather authority and tend to oppress performances by cramping the very IOM's out of which they are built (Newman, 1969); this despite our growing realization that no expression of standards can bind performance by pressing pursuit of its authority. Indeed, it is a painful paradox that lasting authority depends on measured mobilization of resistance to mounting explicitness.

THE LABORATORY

The confounding subtlety of IOM's resides in the subdued declarativeness of their operation. Rather than commanding compliance, they engage us by *framing* performance, a strategy of *parsimony* and not elaboration. IOM's stand at the forefront of the "less is more" paradox, a dynamic best documented in children's acquisition of language (Elman et al., 1996). Because they function by internalization to facilitate creative performances, their robustness requires a demanding (though not rigid) economy, suitable for achieving balance in the personalization process. As in the case of a native language, they thrive on a degree of institutionalization, but one falling well short of dictation of particular performances. Their authority lies in how well they facilitate those performances, a calculus difficult, if not impossible, to formulate either in advance (e.g., Esperanto) or to reconstruct after the fact (e.g., modeling in cognitive linguistics). There is an inevitable dissonance between the delicate informational dynamics suitable to the demands of performance framing, and the degree of articulation associated with more elaborate *a priori* or *post facto* accounts of performance production.

Under the influence of a robust IOM, a performance context, both individually and socially, is rendered a laboratory in which felicitous balances between constraint and freedom ARE *perpetually sought in the rendering process.* No laboratory independent of the performance-generation process itself has yet proved adequate to replicate the exquisiteness of these self-organizing experimental conditions. The few efforts enjoying limited success in synthesizing these conditions can be read to suggest the broad outline of the dynamic I am describing (Elman et al., 1996).

The inevitable and stalwart attempts to generate elaboration of standards beyond the robust residue of these natural framing experiments are integral to these very experiments. Enormously challenging to investigators is the long coexistence side-by-side of low profile but enduring minimal frameworks and high profile over-articulated dogma, both generated in the same experimental process. The authority asserted on behalf of the dogma thinly veils the force invoked to authenticate the more elaborated institutionalization and its ultimate futility.

It is in the semi-deliberate process of personalization that the balance between more and less automated ingredients of performance is and *must be* struck. The complexity of the time scales within which this fine-tuning is brought about and their interrelationship bars the replicability of experimental conditions at any significant distance from personalization itself. The abiding sense of artificiality in the laboratory procedures of experimental psychology and cognitive science bares witness to the inescapable function of personalization and its indigenous time ecology as the only context sufficiently rich to adequately carry on these experiments. Discoveries, some very recent, that learning alters the brain in significant ways through pruning and elaboration of neural circuitry in both childhood and adulthood identify the physiological component of these personalization experiments (Demasio, 1994; Edelman, 1987). That the experimental results are "laid down" in this manner tends to confirm the crucial endogenous experimental nature of personalization.

IOM'S AND DOGMA

It is thus made tractable to identify the onset of IOM's both historically and developmentally and to begin to trace the course of their formulation and reformulation in human individual and social evolution. It is also made tractable to account for the peculiarities of the vexed relationship between IOM's and dogma. The contribution of IOM's to discovering balance in personalization experiments is a cognitive harness lying somewhere between guidance and discipline. The sense of oppression in attempts to perform lies on the road to dogma in the absence of adequate guidance on the one hand and in the drift beyond discipline on the other. This roughly identifies a zone of useful constraint on behavior arising out of our engagement with certain things in the world, often of our own making, socio-culturally and/or individually (Clark, 1997; Hutchins, 1996; Norman, 1990). A developing central and sensory-motor system is subjected to a series of such harnesses, evolved by nature or ourselves, measures their utility by increments of mastery, leaves some behind and carries some forward by internalization in memory (Wohlmuth, 1997), and lays down neurocircuitry as a scaffolded residue that promotes replicability (Damasio, 1994; Edelman, 1987).

What languages share throughout the world, with all of their differences, are the ingredients that together provide a harness for shaping linguistic performance within a reasonably comfortable zone of constraint (Deacon, 1997). Only natural experiments crossing evolutionary and developmental time could have mobilized temporal complexity in the manner necessary to establish this zone which is now deeply associated with being human. Language thus serves us not only as an overarching prototype for IOM's but is threaded through all others in a variety of obvious and subtle ways.

All robust IOM's are kindred. They have survived socio-cultural and individual experiments in multiple interrelated timescales stretching from evolutionary time to nanoseconds and in so doing have become deeply insinuated in the energetics of human activity. They are piggybacked on the entrenched status of language in the regulation of that activity. The subtlety of their insinuation disguises their regulative clout so that we are only dimly aware of the dimensions of their grip on our behavior. But the faintness of this recognition contributes rather than detracts from the peculiarity of their power, producing an effect that registers as magic. Aware of this power over us but not its genesis, we consequently are inclined to seek and regularly misidentify its source and engage in prolonged forays of largely futile replication and proliferation often characterized by obsession and mania. Because the stakes in these campaigns are ultimately social control, they can be waged with great ferocity, conscript enormous resources, and exact catastrophic costs.

These often-costly efforts at conscriptions of IOM's prove futile because they enlist points of engagement falling outside the zone of useful constraint, i.e., the results of such efforts prove less than engaging to human minds. To endure, the candidacy of IOM's must be promoted and their utility sifted in the experimental space-time human beings operate within. Indeed, their only impetus and measure of success is that provided relentlessly by the experimental context--generation and survival to engage human activity through mobilization of human resources. That devotion to ritual practices attracts armies of adherents, much more than the reverse, suggests that the residues of personalization experiments gather momentum in both long and short time scales. It should come as no surprise that armies are more likely to undermine the devotion they champion than sustain it (except perhaps devotion to army discipline), and lose their *raison d'être* to serve personalization in the process.

While challenging, the generation of IOM's in ritual is easier to grasp, as is the undermining of ritual devotion by efforts at proliferation, than the more secular analogue of this process: the relationship between IOM's and canonical texts. IOM after IOM has been subject to elaboration and proliferation beyond the zone of useful constraint. The driver's handbook ubiquitously used in training, for example, while seemingly derivative of the Motor Vehicle Code, has its genesis in common sense under articulated "rules of the road" which have been *elaborated* in the Code and the handbooks in use today. Characteristic of IOM's, the handbook, unlike the Motor Vehicle Code, has been subject not only to elaboration, but pruning/tightening as well, in bringing it more squarely into the useful-constraint zone (Wohlmuth, 1995). This largely (but not entirely) personalization-friendly process conscripts whatever paraphernalia is culturally available. It has waxed multimedia in recent decades with the inclusion of audio-visuals as more parsimonious bearers of information content engaging the dynamics of good driving. The greatest pressure on personalization now comes from the recruitment of ever more powerful graphics, deployed in software, CD's, video packages and interaction training, to enrich the zone of utility and perhaps overhaul it. These developments now begin to impinge on the codification process itself, thus establishing a new relationship between driving IOM's and driving codes, a reorganization likely to be reflected in other domains of public regulation (Wohlmuth, 1999).

Recognition of the role of IOM's in the personalization process sheds light on the historical ascendance and contemporary descendence of canonical texts in a variety of domains and fields. The insight that the actual authority of canonical texts has always been mediated by IOM's is grounded in an adequate understanding of personalization as the gateway to regulation of human behavior. This insight has been gradually forced upon us in the increasingly visible relentlessness of the personalization process itself. Amidst the accumulating clutter of explicit standards for performance, we inevitably find ourselves seeking more reliable, trustworthy and felicitous grounds for judgment and action. Parsimony drives this pursuit, a delicate sustainable economy hammered out moment by moment and day by day by the demands for achieving a viable internal-external balance in the formulation of our behavior. While these demands do not guarantee the quality of resulting performances, they are the *only* constraints ultimately available for sculpting the zone of utility, resting as they do on our evolutionary heritage in general and our linguistic heritage in particular.

SCIENCE, POLITICS AND ART

While the quality of the rolling outcomes of the personalization process cannot be guaranteed, we do have a limited capacity to help shape the laboratory in which IOM's are forged, a laboratory whose momentum rides on the creative energy released by successfully cultivating the constraint zone itself. Whatever power we have to individually and socially intervene in our destinies is inextricably linked to this cultivation process.

In our own time it has become increasingly possible to acquire a perspective on the dimensions of the personalization process, however limited our means for intervention. This enriched point of view has been born out of the futility of canonically hijacking that process. But engaging dogma willy-nilly can yield *only* glimpses of this phenomenon. Transitional tools providing access to IOM's are necessary to give muscle to observations that may ground interventions. Such tools, diverse though they may be, share a capacity to penetrate the apparent authority of dogma to discover and revitalize the points of personal engagement lying underneath--the robust "hypotexts" of ongoing practices.

To remain robust, every viable discipline and practice must have such tools, though the authority invested in the currency of dogma minimizes their profile. In hard science, they lie in the acid of day-to-day experiment and replicability rendering always provisional the authority of entextualized scientific truth. Thus the heartland of scientific practice is an agnostic methodology that, at its best, recruits dogma (hypothesis and theory) in an institutionalized personalization process. While hard science, like all other human activities, tends to be oppressed by its extant dogma, it internalizes the means for relief from oppression at critical thresholds of tyranny.

In social and political practices, transitional tools, while available, are only ambivalently institutionalized, and are consequently less well-developed and more tenuous than in hard science. It is often acknowledged/observed that healthy socio-political structures thrive on openness, but what this openness amounts to and how it is achieved tends to be expressed dogmatically, whether in democratic or undemocratic terms. The assertion of dogma when this issue is pressed belies fundamental doubt as to what the actual nature and source of openness is and, whatever it is, how compatible it can be with adequate stability. The demonstrated difficulty of transporting felicitous openness from one polity to another strongly suggests the illusiveness (or non-existence) of a formula. Institutionalized dogma does not speak for itself--its redeemable heartland lies in the hypotext of IOM's, and transitional tools are mandatory to bridge the two. Unlike the scientific type, socio-political dogma resists recruitment for productive personalization experiments.

I conclude by observing that the arts, not the sciences, embody the fundamental grounds for transition in all of these systems, including science itself. Art serves culture by institutionalizing the means of transition between dogma and IOM's. Amidst the oppressiveness of dogma, the exhilaration of artistic transmission reaches deep inside the personalization process and motivates it (Arnheim, 1971, 1996). That social change is an experimental process is suddenly revealed and greeted with characteristic human ambivalence. We can become, at times painfully, aware in our creative moments why sonata form, blank verse, and non-Euclidian geometries are capable of engendering reverie and reverence simultaneously and, in the day-to-day muddle of our lives, alas indistinguishably. The controversiality of art and the tenuousness of our commitments to its flourishing betray our ambivalence toward the creative implications of the centrality of personalization. Only courageous confrontation of the sources of that ambivalence is likely to safeguard the critical role of art in the dynamics of individual and social growth.

REFERENCES

Arnheim, R., 1971, *Entropy and Art: An Essay on Disorder and Order*, University of California Press, Berkeley.

Arnheim, R., 1996, *The Split and the Structure*, University of California Press, Berkeley.

Clark, A., 1997, *Being There: Putting Brain, Body, and World Together Again*, MIT Press, Cambridge, Mass.

Damasio, A. R., 1994, *Descartes' Error*, G.P. Putnam's Sons, New York.

Deacon, T. W., 1997, *The Symbolic Species: The Co-evolution of Language and the Brain*, W. W. Norton & Co., New York.

Edelman, G. M., 1987, *Neural Darwinism: The Theory of Neuronal Group Selection*, Basic Books, New York.

Elman, J. L. et al., 1996, *Rethinking Innateness: A Connectionist Perspective on Development*, MIT Press, Cambridge, Mass.

Hutchins, E., 1996, *Cognition in the Wild*, MIT Press, Cambridge, Mass.

Newman, W. S., 1969, *The Sonata Since Beethoven*, University of North Carolina Press, Chapel Hill.

Newman, W. S., 1972b, *The Sonata in the Classic Era*, 2nd ed., W. W. Norton & Co., New York.

Norman, D. A., 1990, *The Design of Everyday Things*, Basic Books, New York.

Wohlmuth, P. C., 1995, Traveling the highway: sources of momentum in behavioral regulation, *J. Contemp. Legal Issues* 6:1.

Wohlmuth, P. C., 1997, Jurisprudence and memory research, *J. Contemp. Legal Issues* 8:249.

Wohlmuth, P. C., 1999, *Infrastructure* (video), Inst. for Law & Systems Research, San Diego.

UNDERSTANDING IMPERFECTION

Martin Zwick

Professor, Systems Science Ph.D. Program
Portland State University,
Portland OR 97207-0751 USA

INTRODUCTION

In this paper, I want to present a conception that I have been working on for a number of years (Zwick, 1983, 1995) about the use of systems ideas. I start from the negative and proceed to the positive. The negative assertion is that systems ideas by themselves, unsupplemented by more specific and concrete knowledge (e.g., from the various disciplines), are insufficient for practical application, either for obtaining knowledge about the world or for solving problems. The reason for this is the inherent abstractness of these ideas. The positive assertion is that this metaphysics can be a rich source of insights needed to improve the human condition and a component of a new scientific worldview.

IDEAS ARE FOR INSIGHT

The abstractness of systems ideas is succinctly captured by Bunge's characterization (1973) of the systems project as the attempt to construct an "exact and scientific metaphysics" (ESM). By *metaphysics* Bunge means general propositions about the world which hold for a wide variety of systems ("metaphysics" here thus does *not* refer to questions of the existence of God, free will, etc.). By *scientific*, he means grounded in, i.e., drawing upon and contributing to, the sciences. By *exact*, he means mathematical, or capable ultimately of being expressed mathematically (Bunge actually includes "exactness" within "scientific," but I pull it out as a separate idea). This ESM is presently constituted by a *multiplicity* of systems theories, e.g., information theory, control theory, game theory, and the like. A *singular* systems theory does not exist. In a sense, this is the goal of the systems project, a systems "theory of everything" (TOE) radically different from the more familiar

Understanding Complexity, Edited by Ragsdell and Wilby.
Kluwer Academic/Plenum Publishers, New York, 2001

TOE sought by physicists trying to unify the four fundamental forces. A systems "theory of everything" would be a true metaphysics, which would pertain to everything, not just the physical world and not merely in principle; it would touch every discipline, not merely physics.

Here is a slightly different conceptualization: systems theories are less abstract than mathematics and philosophy but more abstract than theories in particular scientific disciplines (Boulding, 1956). By being more abstract than specific theories in the natural and social sciences, systems theories cannot without supplementary knowledge be evaluated empirically or used for practical purposes. Actually, even scientific theories are not directly testable. Theories, applied to different phenomena, offer models of these phenomena, and models, not their parent theories, are confronted by empirical evidence. Models are linked sets of relations, each a hypothesis which can be falsified, but models are not necessarily rejected by the failure of specific hypotheses; they may be rescued by plausible *ad hoc* adjustments. Finally, relations involve observables which define a "model object" appropriate to the phenomenon under study. This epistemological framework, based on Bunge's analysis, is summarized below in Table 1.

Bunge does not include (2) as a distinct level, and I offer simpler names for the other levels. I add in italics the conception that systems theories are triadic in character (ESM), midway in abstraction between E & M and S.

Table 1. Epistemological hierarchy. [Bunge's terms are in brackets].

	mathematics (E) *philosophy (M)*
5. systems theory ["generic semi-interpreted theory"]	*systems theories (ESM)*
4. theory ["general theory"]	*theories in the natural and social sciences (S)*
3. model ["specific theory"]	
2. relation, law, hypothesis	
1. observables ["model object"]	

Models of phenomena (3) are produced either inductively from below by integrating observed regularities (2) or deductively from above by applying a theory (4) to a model object (1). Similarly (in italics), systems theories are produced inductively from below by identifying isomorphisms between different scientific theories (S) or deductively from above by adding interpretations to mathematical formalisms (E) or concreteness (and exactness) to philosophical generalities (M).

This epistemological hierarchy suggests that unless systems-theoretic notions are used in specific scientific theories and more explicitly in models of particular phenomena, they are too far removed from the real-world to offer usable knowledge by themselves. For example, game theory does not explain behavior observed in auctions and cannot guide rational action in such settings, but if supplemented by additional information about economic systems in general and about auctions in particular, game theory could provide the basis for a descriptive and/or normative theory of auction behavior.

However, while an economist might draw upon game-theory to develop a model of auctions, such a model is unlikely to be comprehensible to scientists in other fields, to other scholars, and to the public at large. For the general public, the scientific study of particular phenomena is rarely feasible. It might seem therefore that the project of constructing an "exact and scientific metaphysics" is irrelevant except to scientists.

This is not so. What non-specialists need to navigate the many worldly contexts which they encounter is *insight*. Insight is more usable than theory or methodology because it is flexible and open-ended. Insight is what a metaphysics can offer. The significance of the systems project, aimed at a metaphysics which is precise and grounded, may well consist more in its general cultural impact than its scientific fecundity. This is the positive aspect of the deep abstraction which characterizes systems ideas.

INSIGHT IS NEEDED INTO IMPERFECTION

What is not problematic does not require understanding. It is imperfection which generates the need for insight. In the social world, "imperfection" means war, injustice, exploitation, and the like. Speaking more narrowly about culture, our values are imperfect not merely in the difficulty of realizing them, but because the values themselves are contradictory. This is the anti-utopian position of Isaiah Berlin (1991) who argues that cultural values do not resemble the mutually consistent axioms of a formal mathematical system, but are typically incompatible. In the natural world, "imperfection" encompasses the dangers arising in the environment to human beings and other organisms and, reciprocally, the local and global threats to the environment posed by human action. In the biological world, "imperfection" includes disease, suffering, and (from the vantage point of prey) predation. In the psychological and behavioral worlds, there are imperfections in our characters and our actions. And so on. One need not belabor the obvious. Imperfection abounds in the natural and social worlds. In so far as we must deal with it, we need insight into its deepest, "metaphysical," basis.

I am not speaking of imperfection in our *models* of the world, but in the world itself. That is, I want to discuss ontology and not epistemology, and employ an objective rather than a subjective mode of discourse. Imperfections in models can be included while retaining an objective stance by considering deficiencies in the "modeling subsystems" of organisms, organizations, and societies.

To say that imperfection is objective is not to deny that imperfection implies the vantage point of *agents* who have *interests*. To a first approximation, we humans are the reference point, as we are the ones making judgments. But the reference point is actually larger and includes all forms of life. Every living system has interests and in so far as such a system can directly or indirectly further its interests, it exhibits agency. To make the notion of "interests" exact, one can define for a system a utility function. Systems having interests is the subject of game theory, in which utility is given formal treatment (von Neumann & Morgenstern, 1944). Utility with respect to reproduction is "fitness," hence the use of game-theory in evolutionary biology, but utility is a broader notion. From a biological perspective, the title of this essay might have referred to understanding "vulnerability" or "hazard," but "imperfection" is plainer and more general.

From the perspective of individual living systems, the universe is less than perfect. Instead of speaking of imperfection, one could speak of "problems" which we or other organisms face, but the word "imperfection" has the virtue of suggesting an objective character to these problems. Imperfection is not merely in the eye of the beholder. Problems reflect a mismatch between what is actual and what is ideal to some agent, and what is ideal may be subjective, but it may alternatively be objective. The survival needs of organisms are objective facts. More importantly, the causes of the mismatch between ideal and actual are objective features of the world, and many of these features are ubiquitous, perhaps even universal. This is a Gnostic or Kabbalist view. Imperfection is pervasive, and an exact and scientific metaphysics would lay bare its most general, i.e., ontological, foundations. To regard imperfection as a metaphysical condition, however, does not mean that this condition is irremediable, just that problems are lawful and ubiquitous. Perfection is precluded by limitation, and limitation is the precondition of existence. What remains possible is perfect*ing*, the verb, which is better anyway than perfect*ion*, the noun, a state which is unattainable, because only the process and not the state produces the drama of history, the absence of whose performance would surely be an imperfection.

I want to make several qualifying remarks concerning this point of view. First, let me mention the position, more popular in the past, articulated for example by Leibniz and mocked by Voltaire, that everything is perfect just as it is. It is hard to imagine a 21st century writer who would find such a position still worthy of parody. Even were this true at some "high level of reality," at the level we live on, it is untrue. Moreover, even the highest truth is not the whole truth, and the whole takes precedence over the highest. From a religious, or, more precisely, a doctrinal perspective, the pervasiveness of imperfection raises the theological problem of theodicy, i.e., the conundrum of how there could be a God who is all-knowing, all-powerful, and all-good, given the evil and suffering in the world. A systems metaphysics, being necessarily silent about the existence of God, cannot solve this conundrum, but the "secular theodicy" which such a metaphysics can offer might be a valuable corrective to the Leibnizian optimism of current anthropic speculation.

Finally, I want to acknowledge the negativity which afflicts a focus on imperfection. A glass half empty is also half-full, so the word "imperfection" is itself imperfect in its exclusively negative connotation. Imperfection necessarily encompasses the possibility of perfecting. I mean actually to assert something more like a mixture of perfection and imperfection or a delicate balance between the two, perhaps something like the idea of the "edge of chaos" (Langton, 1992).

SYSTEMS IDEAS ABOUT IMPERFECTION

There are many systems ideas and themes which bear on the subject of imperfection and what follows is only a sample.

- Tensions, not restricted to formal mathematical systems, between incompleteness and inconsistency
- Imbalances involving other fundamental dualities, such as variety and constraint, order and disorder, unity and multiplicity
- Problems of boundary, e.g., the issue of crisp vs. fuzzy boundaries, excessive or inappropriate openness or closedness

- Instability, the destruction of order through chaos or catastrophe; or, conversely: overstability, rigidification
- Pathologies of (feedback/feedforward) control
- Aberrations of hierarchical order, including the opposing needs of centralization and decentralization, differentiation and integration
- Informational parasitism (internal or external)
- Dependence upon and constraint by the environment
- Limitations of agency: multiple deciders or objectives (inconsistency) or uncertainty (incompleteness), problems of global optimization, computational intractability, and undecidability; unintended consequences and counter-intuitive effects
- Difficulties presented by other systems; competition, predation, exploitation; tensions between autonomy and interdependence; dilemmas involving multiple agents
- Embeddedness in and thus control by more encompassing systems
- Impermanence and fragility of order; vulnerability to events on smaller, larger, or the same scale
- Limitations of modeling subsystems: the inherent indistinguishability of variety and noise, of reality and illusion

This is just a list of systems ideas relevant to the theme of imperfection. If one tabulated, instead of ideas, the theories which bear on this subject, one would include most of the well-developed theories central to the systems project, e.g., information theory, graph theory, dynamic systems theory, control theory, game theory, non-equilibrium thermodynamics, evolutionary theory, etc. For convenience in this essay, I use the name "Problematics" as a working label for the enterprise of collecting, formulating, and organizing systems ideas about imperfection. In earlier work (Zwick, 1983), I tried to weave together such ideas into a metaphysical narrative about imperfection. This essay and its precurser (Zwick, 1995) explain the motivation for this undertaking.

The ideas on this list, and other related ideas, can be given exact expression and linked to current scientific knowledge. There is thus the possibility of an ontology of problems, a systems metaphysics organized around the theme of imperfection. But here we come face to face with the imperfection of systems theory itself. As already noted, a unitary theory does not exist; instead we have a multiplicity of theories, weakly linked to one another. Some means must still be found to bring order and unity to this multiplicity.

I suspect what is needed for this is something *like* the hierarchy of systems types sketched by Boulding (1956): (i) frameworks, (ii) clockworks, (iii) control (cybernetic) systems, (iv) open systems self-maintained by matter-energy throughput, (v) genetic-societal systems (plants), (vi) purposeful information-processing systems (animals), (vii) self-conscious symbol-processing systems (humans), and, not at all discussed by Boulding, (ix) "transcendental" systems. Such a hierarchy brings the contemporary idea of "complexity" into proximity with the traditional idea of "being." The notion of complexity, while capable of being interpreted in many different ways, can be made exact and scientific, but the notion of being is metaphysically more substantial. From the perspective of the construction of an exact and scientific metaphysics, it is interesting and probably not accidental that Boulding's hierarchy evokes -- even resembles -- the religious-philosophical idea of the "great chain of being" (Lovejoy, 1936), which was widely accepted in Western culture though most of its history. (This traditional idea also directly addressed the ubiquity of imperfection, viewing it as a necessary part of a greater perfection.) With Boulding's proposal as a first approximation, one can at least imagine a systems-theoretic version of "the great chain" which would satisfy the dual requirements of being exact and scientific. No doubt some of the assumptions (e.g., the principles of continuity and plenitude) which characterized the traditional idea would have to be discarded or altered. As the "skeleton" of an ESM, a modern "chain of complexity" would probably have to be discrete with a modest number of levels, and would have to accept contingency, anathema to traditional doctrine, as fundamentally implicated in the creative generativity of nature.

Why would one undertake to construct a systems-theoretic ontology of imperfection? The motivation is deeper than the mere solution of problems. Metaphysics is a basis for morals (Murdoch, 1992). Spinoza for example begins his *Ethics* (1677) with a systems metaphysics (Jonas, 1965). This approach was natural in the premodern era: one begins with a description of the world and then considers how one should live in it; one moves from "is" to "ought." Philosophers since Hume have however pointed out that one cannot *derive* "ought" from "is." True enough: fact by itself is not sufficient, but it is necessary. Though domains of fact and value may be sharply distinguished, they interpenetrate in a complex manner, like two basins of attraction of a nonlinear dynamic system whose separating boundary is fractal. Moreover, value *emerges* in a world of fact. The existence of value is a fact. The notion of utility, developed in game theory and grounded in biology and the social sciences, bridges the two domains. (More generally, game theory itself, being capable of both descriptive and normative use, exhibits the interconnection of fact and value.) The formal idea of utility is much richer than its colloquial connotation of usefulness, which is largely external in character: utility might also quantify value internal to a system, i.e., perfection of structure. The notion of utility augments those of matter, energy, and information as a fourth fundamental category, instantiated and visible in the phenomena of life. It is a notion which can encompass both the objective and the subjective, both the necessary and the contingent; on the human scale, both needs and wants.

Without understanding what is, one really cannot consider what ought to be, at least not effectively, so metaphysics bears significantly on ethics. Game theory, for example, brings exactness and empirical knowledge into contact with ethics. From insight into the ontology of imperfection there follows the possibility and thus the responsibility of perfecting, of "tikkun," to use a term from Kabbalist metaphysics meaning the *fixing* of metaphysical imperfection as far as is possible. Or, as the philosopher Levinas (1989) has argued, one should even reverse the sequence and say that ethics *precedes* ontology because it motivates -- or should motivate -- metaphysics, as it does for Spinoza.

"Problematics," that is, systems inquiry into the origins and nature of imperfection is not sufficient. Problematics is less than "diagnostics" -- being able to say what the critical issues are in any particular problem. One might imagine a physician's desk manual (a DSM-V) for the diagnosis of metaphysical ills. Diagnostics in turn falls short of "therapeutics" -- knowledge of what to do about the problem, of how perfecting might be accomplished. But the progression from problematics to diagnostics to therapeutics, however important, is not to be rushed. Facile diagnostics and simplistic therapeutics are often counterproductive, this being one of the lessons gained from systems thinking. The assumption that we already understand problems adequately is usually illusory. What is needed most to solve problems is not theory or methodology but insight, if it is deep and especially if it is widely shared.

THE CONTRIBUTION OF SYSTEMS IDEAS

A metaphysics of imperfection would contribute to a new scientific world view which is coherent and moral. In a reductionist world view, there is no imperfection, only fundamental particles and their interactions, and life is a mere epiphenomenon. Systems thought offers a radically different view. It replaces the hegemony of the fundamental with the priority of the central, and hence of the general. From the perspective of the central, all systems are ontologically equal (though from the perspective of "complexity" -- or "being" -- they are not). Ultimately, also, all are flawed by limitation, which is the price of existence. To use Murdoch's phrase (1992), systems are "*partial* wholes" [emphasis added]. Again, it is important to point to the virtue in the flaw. In the words of the poet Leonard Cohen, "there is a crack in everything; this is to let the light in."

A systems metaphysics can be a bridge from science to the humanities, the arts, philosophy, religion. There was once, from the medieval period to the beginnings of modern science, an integrated view of matter, life, mind, cosmos. This view suffered, however, from an overvaluation of order and harmony, from being inexact and empirically undeveloped, and from being tied to one religious tradition. It eventually disintegrated with the rise of science, leaving a cultural vacuum described poignantly in Yeats' complaint, "...all coherence lost...". Post-modern nihilism may be the final act of this decline, but other developments hint at the possibility of a recovery of cultural coherence, of a new integrated world view centered in science (Toulmin, 1982) but open to other realms of human experience. The project of constructing an exact and scientific metaphysics, however fragmented and fragile an undertaking, is a vital part of this possibility.

It is significant also that systems ideas can contribute not only to culture at large but to "personal knowledge" (Polanyi, 1964), which can be individually appropriated and used in our private lives. Because the systems view does not reduce, because it gives priority to form (which we perceive) over matter (about which our direct experience is limited), because it seeks cross-level truths (applicable "above" to natural and social systems, and "below" to ourselves), and because it is not committed solely to 3rd person objectivity but encompasses also 1st person subjectivity, for these and other reasons, systems ideas, unlike most ideas in science, can be meaningful to us. They might even help us understand the lawfulness but not hopelessness of our own imperfections.

To close the circle of this argument (and perhaps return to the negative): the strength of an exact and scientific metaphysics is also its weakness. The main obstacle preventing systems ideas from contributing significantly to a new scientific world view and to personal knowledge is the abstractness of these ideas, the very abstractness from which their scope and significance derives. This abstractness limits the number of people who become aware of systems ideas. It also tends to restrict the understanding of these ideas to a strictly intellectual mode, which lacks the emotional force needed to change our image of and relation to the world. This poses an important educational challenge, but that is another subject.

REFERENCES

Berlin, I., 1991, edited by H. Hardy, *The Crooked Timber of Humanity*, Alfred A. Knopf, New York.

Boulding, K., 1956, General Systems Theory - the skeleton of science, *Management Science* 2:197-208.

Bunge, M., 1973, *Method, Model, and Matter,* D. Reidel, Boston.

Jonas, H., 1965, Spinoza and the theory of the organism, *Journal of the History of Philosophy*, 3:43-58.

Langton, C., 1992, Life at the edge of chaos in: *Artificial Life II. SFI Studies in the Sciences of Complexity, vol. X*, by C.G. Langton, C. Taylor, J.D. Farmer, & S. Rassmussen, eds., Addison-Wesley, Redwood City, CA

Levinas, E., 1989, Is ontology fundamental?, *Philosophy Today*, Summer 1989, p.127.

Lovejoy, A., 1936, *The Great Chain of Being*, Cambridge: Harvard University Press, Cambridge.

Murdoch, I., 1992, *Metaphysics As a Guide for Morals*, Penguin, London.

von Neumann, J., & Morgenstern, O., 1944, *Theory of Games and Economic Behavior*, Wiley, New York.

Polanyi, M., 1964, *Personal Knowledge*, Harper and Row, New York

Spinoza, B., 1677, translated by A. Boyle, *Ethics*, Dutton, New York.

Toulmin, S., 1982, *The Return to Cosmology,* University of California Press, Berkley.

Zwick, M., 1983, Incompleteness, negation, hazard: on the precariousness of systems *in: The Relation Between Major World Problems and Systems Learning, Vol. 1, Proceedings of the 1983 Annual Conference of the Society for General Systems Research*, Intersystems, California.

Zwick, M., 1995, Towards an ontology of problems, *Advances in Systems Science and Applications* Special Issue I:37-42.

ENGAGING WITH TRANSIENT COMPLEXITY IN DEVELOPMENT PROJECTS

John Friend

Centre for Applied Development Studies
University of Lincolnshire and Humberside
Lincoln LN6 7TS, UK

INTRODUCTION

This paper reflects the perspective of a researcher and process facilitator whose endeavours to engage with complexity over four decades have roots not so much in the systems movement as in three other parallel movements or ideologies – those of operational research, action research and participatory development. So the question I shall explore will be whether systems science need be seen as the only starting point - or indeed always the most helpful one - in engaging with complexity in the world around us.

I shall argue that there is a considerable potential for synergy between systems ideas and less synoptic approaches that have evolved to address complexities of a more transient nature. For the first time in my career, I now find myself in an academic environment populated by scientists who are well versed in the various schools of thought within the systems movement. This, together with the opportunity I have been given to contribute to this volume, I see as presenting exciting new opportunities for me to share perspectives with systems scientists engaged in our common endeavour of understanding and managing complexity in a volatile and politically fragmented world.

First, I shall introduce some concepts that I have found useful in my own research and consulting career. This will lead me into a review of some of the prospects I see for future synthesis, with particular reference to the management of international development projects.

Understanding Complexity, Edited by Ragsdell and Wilby,
Kluwer Academic/Plenum Publishers, New York, 2001

VOCABULARIES FOR ENGAGING WITH COMPLEXITY

Among the writers on systems thinking who are most widely recommended to students of management in my country are Checkland (1981) and Flood and Jackson (1991) – writers who would place themselves at the softer, in the sense of less quantitative, end of the systems spectrum. These authors recognise a wide diversity of systems-centred approaches to the study of organisations, yet have sought to identify a core of shared beliefs in such fundamental systems concepts as holism, hierarchy, emergent properties, communication and control.

In my own research and my own practice, I tend to refer primarily to less synoptic and more decision-oriented terms such as planning, negotiation and facilitation. There are of course some overlaps between these vocabularies; and one of these lies in the use of the term complexity itself. Yet I have found that such terms as *complexity* and *problem solving* may carry subtly different meanings for those of us who draw on different traditions of thought.

Many of the forms of complexity that have concerned me in my applied project work seem to have been transient or ephemeral rather than systemic in structure. So my contention is that they are less amenable to systems approaches than to other traditions of scientific thought. I have frequently found myself speculating on the existence in several languages (the best-known being Spanish) of two subtly different translations of the English verb *to be*. The verb *ser* in Spanish indicates a relatively durable form of existence – I *am* a scientist; the most serious problem in this country *is* that of poverty. The alternative verb *estar* indicates a more transient state of being – I *am* tired; the main problem in this country *is* (at this time) one of accommodating refugees from its recent natural disaster or civil war.

THE EVOLUTION OF A DECISION-CENTRED APPROACH TO COMPLEXITY

As a foundation for further exploration of such issues, I should first say more about the main contextual and ideological influences to which I have been exposed in the course of my working life. In ideological terms, I have found myself successively exposed to, and influenced by, three movements each of which clearly has links with the systems movement, yet seems to me to merit recognition in its own right:

The **operational research movement** (operations research in the USA) – with particular reference to an emergent "soft OR" tendency in UK.

The **action research movement** - with particular reference to the tradition associated with the Tavistock Institute of Human Relations in the UK.

The **participatory development movement**, emerging from convergent trends in such fields as agricultural extension, organisational development and community education.

I see it as significant that none of these movements is primarily academic in its roots. In all of them – and here I also include the systems movement - it is significant that there have been gradual shifts of emphasis that have brought increasing flexibility of application. Yet some of the labels that have been used to describe these shifts (such as *soft operational research* or *critical systems thinking*) carry more meaning to the scientific communities from which they come than to any lay people with whom these scientists may engage.

Just as important to me as these different ideologies have been the differences of context in which I have found myself working over almost five decades, alongside managers, planners and other practitioners with widely differing views of the world around them. So let me now reflect further on the journey that I have made during my own career as a management scientist and facilitator in an increasingly wide range of organisational, inter-organisational and community settings, recognising that many of us have made similar journeys over the course of our working lives.

I shall divide my journey broadly into five decades, from each of which I have been able to draw different lessons.

The Fifties: Exposure to Industrial Management and Operational Research

After graduating in mathematics in 1952, I was keen to escape from the rigours of the academic world. I then found myself starting to engage with problems of industrial and commercial management through successive statistical and operational research positions in the steel, airline and chemical businesses. During this period, I continued to spend most of my time analysing numerical data – time series data in particular – and working on mathematical models to generate predictive power in relation to replicable operations within a clearly-structured corporate context.

In 1958, my airline employer sent me to the United States to attend an introductory course on operational research led by Ackoff and Churchman in Cleveland. It was only then that I began to embrace the view of OR as something broader and more exciting than the application of statistical methods in a business context. Indeed, it was during this short course that I was first introduced to the ideology of systems thinking and became aware of the power of such terms as holistic, proactive and stakeholder – all of which were fresh to me at the time.

It was not however until the sixties that I was to encounter influences that led me towards a painful yet ultimately liberating transformation in my entire research orientation.

The Sixties: Local Government, the Social Sciences and Strategic Choice

The most significant turning point in my career came in 1964 when I joined a new *Institute for Operational Research,* set up as a semi-autonomous unit within the matrix of the Tavistock Institute in London. Here I worked full-time on a four-year interdisciplinary study of policy making in Coventry City Council, and found myself experiencing a dual culture shock. On the one hand, I faced the challenge of making sense of an organisational structure in which power was widely distributed among politicians, administrators and professionals, within a culture of collective decision-making through committees and informal political groups. At the same time, I found myself having to learn to work alongside social scientists as colleagues, with very different ideas to my own about the process of research.

With my project colleague – a social anthropologist - I was encouraged to spend much of my time attending committee, departmental and political meetings as a participant observer, and talking informally with individual officials and politicians. Over three years, we attempted to trace the course of many interrelated planning issues as they were steered through this complex machinery of decision-making. What we saw was that commitments to actions and policies tended to be made incrementally, alongside steps to deal with residual obstacles through various types of exploratory action – which included not only investigations but also policy consultations and negotiations with people working on other related agendas.

From this experience, a view of strategic decision-making emerged that was far from a holistic one. We could no longer take a conventional view of planning as a top-down process, starting with agreement on high-level organisational goals. More realistically, we began to see it as a continuous and diffuse activity of commitment-building in the face of multiple sources of *uncertainty* – with recurrent disagreements as to their relative importance and the actions to be set in train in response.

We saw both the agendas of choice and the configurations of participants as continually changing in response to fast-moving events. This kind of activity we saw as driven primarily by political pressures to take decisions of some urgency and importance. Yet the more complex the connections among the issues, the more daunting became the uncertainties to be addressed. So the more prominent became the challenge of *managing uncertainty in a strategic way.* Indeed, the more complex the issues, the more prominent became the management of uncertainty as a *secondary driver of progress in decision making* - secondary only to the taking of urgent decisions. We found ourselves expressing the essence of this view in terms of a conceptual model of three broad types of uncertainty, as follows:

- Uncertainties in the working **E**nvironment (**UE**), calling for a *technical* response;
- Uncertainties about guiding **V**alues (**UV**), calling for a more *political* response;
- Uncertainties about Choices on **R**elated agendas (**UR**), calling for a *negotiative* response.

This tripartite classification of sources of uncertainty in decision-making became the core of a *decision-centred view of planning* (Faludi 1987), differing radically from the system-centred models of planning that were becoming fashionable during the late sixties in government and in city planning, as in the world of business. This decision-centred view has now stood the test of time. Upon it has been built an interactive planning methodology known as the *strategic choice approach* (Friend and Hickling, 1997) that has since demonstrated its relevance to a wide range of organisational, inter-organisational and community settings.

The Seventies: Inter-Agency Planning, Political Science and Workshop Facilitation

The seventies saw a succession of further Tavistock research and advisory projects addressing a variety of interrelated fields of public policy-making – physical, social, economic and environmental. Among the first such projects was a further longitudinal study in which the focus was not on decision-making within a single corporate organisation, but on the processes of *inter-organisational* planning that my colleagues and I had found to be pervasive in a variety of public policy arenas (Stringer, 1967).

The subject of this research project was the planned expansion of the small country town of Droitwich to absorb population from the war-damaged city of Birmingham. The two organisational partners in this development programme were both elected local authorities, one representing the 6000 residents of the town itself, and the other the 500,000 residents of the surrounding county of Worcestershire. A joint development committee was formed with eight members appointed from each side, despite the disparities in their resources - and also in their stakes in the intended quadrupling of the population of the town.

The earlier research in Coventry had been launched at a high-profile event where the city's policy-makers had bestowed their full corporate blessing on the project. In Droitwich, by contrast, agreement to provide access to the researchers was slipped through quietly at a meeting of the joint committee with little understanding of what this might mean. From this modest point of entry, it was then up to the researchers to work progressively outwards, making contact step by step with people in other relevant agencies, and negotiating meetings with them as their significance was progressively revealed.

During the most crucial year of the fieldwork, John Power, a visiting Australian political scientist, joined the project team. Together we developed a conceptual model that posited a state of dynamic tension between two complementary types of configuration for decision making in the public domain. The first type was that of the relatively durable *policy system*, built around a set of generic rules within which to deal with some predefined *category* of choices – for example, choices about highway construction or the management of public housing within some recognised administrative territory.

The second type of configuration was that of the relatively transient and fluid *decision network*, formed through negotiation among individuals to address some here-and-now cluster of *specific* choices transcending the categories and rules covered by established policy systems. Power coined the term *reticulist* for those skilled in fashioning and managing such non-formal networks, and became absorbed in the challenge of understanding the dimensions of *reticulist skill.*

During the course of this project, a set of simple graphical mapping conventions was devised to help in understanding the ever-evolving relationships between decision networks and policy systems. Interviews with decision-makers in many roles were used to construct such graphs for many different configurations of here-and-now decisions. Each of these graphs was superimposed on the same organisational base map on which the main policy systems had their own fixed locations. On this base, the more ephemeral decision networks were then superimposed, taking one focal area of decision – as viewed from the perspective of one respondent - at a time (Friend, Power and Yewlett, 1974).

Some of these policy systems could be shown as nested within other policy systems designed to address broader categories of choices. For example, a system dealing with school policy within some administrative area could be seen as nested both within a national policy system for education, and also within a local policy system, with quite different political accountability, covering a broader range of services in that area. This meant that the relations among policy systems could not be represented as exclusively hierarchical in form.

So it was that the device of the Venn diagram, allowing for membership of partially overlapping sets, became an essential tool for mapping relations among policy systems, as well as the more transient relationships between policy systems and decision networks. The former we represented by box-type shapes with straight-line boundaries, and the latter by broken-line boundaries of more fluid shape.

One outcome of such overlaps among policy systems, it was observed, could be a state of acute or incipient *policy stress,* in which conflicting guidelines emanating from different sets of generic rules had somehow to be reconciled at local level. We witnessed the use of various strategies for the relaxation of policy stress in practice – including not only the exercise of informal reticulist skills but also of argumentation about the interpretation of generic policy statements. Such argumentation could lead in turn to the conscious use of semantic ambiguity or *nebulosity* by those responsible for the design of formal policy statements (Friend, 1977).

In parallel to this research, I now became involved with several Tavistock colleagues in applied projects in which we were invited to test in practice the set of methods for structuring communication about complex decisions that had come together during the Coventry project. These methods, collectively known as the *strategic choice approach,* quickly became seen as a promising means of *facilitating* group decision-making, especially in inter-agency settings (Friend and Hickling, 1997). Among the countries in which these pioneering applications were set were not only Britain but also The Netherlands, Canada and Brazil.

The Eighties: Reflecting on Experience in Negotiated Project Engagements

By 1980, the Strategic Choice Approach (SCA) had becoming internationally recognised as a practical and flexible means of helping groups to make progress together in addressing tangled configurations of choices that had defied other more tightly-structured approaches to planning and strategic management (Hickling 1989). Also, later in the eighties, SCA became accepted as one of the leading "soft OR" approaches for tackling complex and ill-structured problems (Rosenhead, 1989). Where it differed from some other "soft" approaches, such as that called *Soft Systems Methodology* (Checkland, 1981), was in not claiming to be based on a framework of systems thinking. Indeed, it appeared to be at its most useful in helping people with different accountabilities and allegiances to come together to address issues of transient rather than systemic complexity – issues that could nevertheless often be of burning current concern to those involved and the wider interests that they represented.

So the Tavistock researchers were now becoming recognised as pioneers in developing participatory methods for supporting complex and ever-changing configurations of choices in ephemeral decision networks, rather than choices of direction in more stable policy systems.

A further development of the early eighties came with a six-month grant from the then UK Social Science Research Council to myself with two of my senior Tavistock colleagues. The aim was for us to review the experience of past action research projects, with particular reference to projects in which we sought to bring together social science and OR skills. So here was a chance for three of us who had engaged in some of the most challenging Tavistock projects of the sixties and seventies - plus one visiting Canadian scientist - to work together to explore the conditions for successful synthesis of different scientific approaches (Friend, Bryant, Cunningham and Luckman, 1998).

Of the set of 13 past projects that we reviewed, most had involved complex and evolving patterns of relationships between people in the following three contrasting types of role:

The project agents – those primarily responsible for carrying out the project work, whether as researchers or consultants, or in some combination of these roles;

The project hosts – those in whose environment the project work is done, whether acting as clients for the work or merely providing access to the project agents;

The project sponsors – those who provide the resources to enable the project work to be carried out, often in pursuit of more generic research or other programmatic aims.

The distinction between project *hosts* and *sponsors* is one that is rarely recognised in conventional models of action research and project management, which tend to assume that these two roles are combined within a "client" or "client system" role. Nevertheless, the distinction between hosts and sponsors had been significant in almost all the projects within our sample of thirteen. Furthermore, in many cases, those participants who came together *within* any one of the three types of project role – agent, host or sponsor – often came from quite disparate backgrounds, with little or no prior history of working together. So it became unrealistic to talk of an agent "policy system", a host "policy system" or even a sponsor "policy system". Within each of the three types of role, a much more transient configuration of relationships might have to be explored.

So as to investigate this type of complexity further, we developed a conceptual model of the changing relationships among people in agent, host and sponsor roles before, during and after the course of what we called a *negotiated project engagement.* In developing this model, shown in Figure 1, we drew extensively on the work of Boothroyd in his book *Articulate Intervention* (1978), in which he explored the practice of operational research. In exploring the variety of *programme influences* on any OR project. Boothroyd in turn had drawn deeply on the writings of modern philosophers of science, including in particular Lakatos (1972).

In Figure 1, all those who engage in the management of a research, consulting or organisational change project in either agent, host or sponsor roles are seen as drawing on three types of *programme strand* – institutional, disciplinary and personal. The more ambitious the project, the less likely it is that those participating in any of the three types of role will share a substantial history of working together. The team, of project agents may not only be multi-disciplinary, and composed of people with diverse personal motivations; it may also be multi-organisational in its allegiances. Those in host roles may have equally diverse disciplinary or cultural backgrounds, personal motivations and institutional allegiances. Then a joint sponsorship structure may add a further layer of complexity to be explored.

The Nineties: International Development and Extended Participatory Methods

So how can people in any of these roles – agents, hosts or sponsors - learn from their cumulative experience in successive "negotiated project engagements" of such diversity and complexity? In the eighties, we had started to address this question by developing a check list of questions with which to explore influences before, during and after a project engagement in terms of each of the various programme strands of Figure 1. Yet for many years we were to have few chances to test and extend this kind of questioning framework.

In the nineties – however, I began to find new applications for this framework for the investigation of transient complexity. Now – no longer in the Tavistock Institute - I began to engage increasingly with the world of *international* development, through visiting roles first at the University of Sheffield and then at the University of Lincolnshire and Humberside.

The institutions and programmes whereby resources from wealthier countries became channelled towards the perceived needs of poorer countries in the so-called developing world are as important and fraught with complexity as any that we face as we enter the twenty-first century. However well intentioned may be the policies that guide international development agencies and banks, or national governments and charitable foundations, the difficulties that arise in negotiating and managing ambitious international projects are widely recognised.

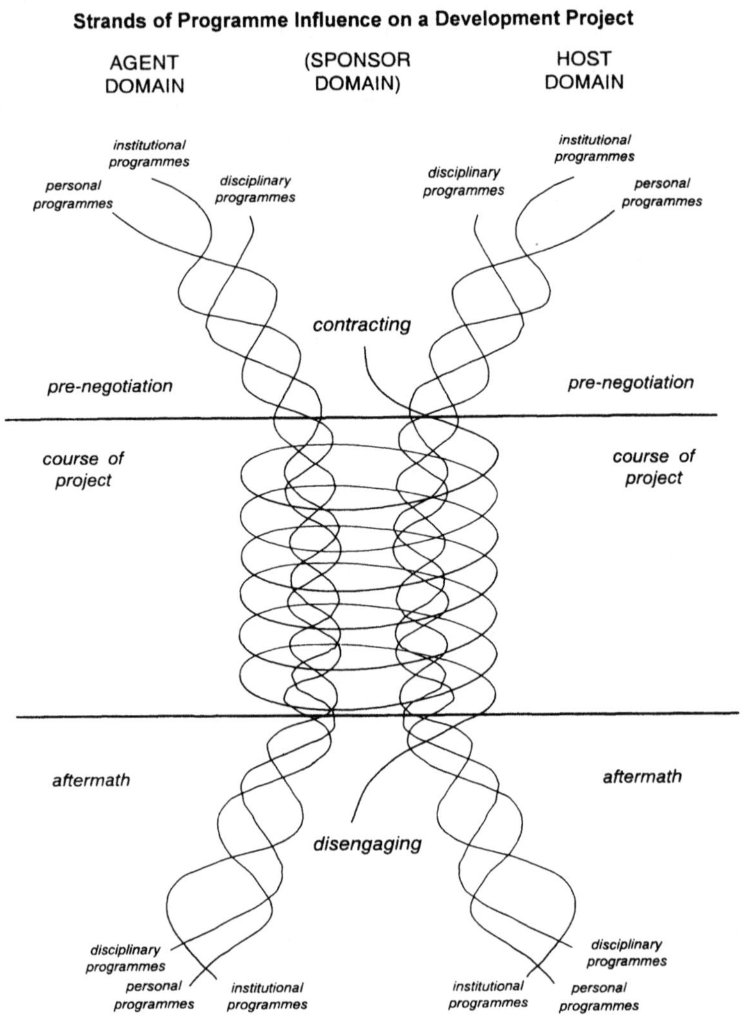

Figure 1. Programme influences on a negotiated project engagement. To avoid excessive complexity in the diagram, the strands relating to those in sponsor roles are not shown.

It was in 1997 that the chance arose for me to join the Lincoln School of Management as a part-time research professor, to take on the challenge of setting up a new Centre for Applied Development Studies. My collaborator in this initiative was my former Sheffield associate Dennis Finlayson - now chair of the new *Applied Systems in Development* Special Integration Group within the International Society for the Systems Sciences. My role in Lincoln has given me further brief but stimulating opportunities to facilitate strategic choice workshops to address pressing issues of local and regional development alongside co-facilitators from developing countries. In several of these workshops, it has been possible for me to introduce software (see the website http://www.btinternet/~stradspan) based on the principles of the strategic choice approach as a tool for recording progress in larger groups and for direct support of smaller groups. Yet the basic workshop approach remains based on interactive visual methods using the appropriate technology of flip charts, pens and walls.

Through this work, I have come to recognise that operational research scientists and systems scientists have much to learn from those who have contributed to the worldwide *participatory development movement.* Significantly, this movement appears to have followed a similar trend over the last two decades, away from the use of formal, quantitative methods and toward a softer, more facilitative approach.

One of the most persuasive writers on participatory development has been Robert Chambers at the University of Sussex. As an alternative to the then dominant paradigm of gathering data on rural communities through large-scale quantitative surveys, he first advocated a more informal approach of *Rapid Rural Appraisal* or RRA (Chambers, 1983); now superseded by a more interactive concept of *participatory rural appraisal* (PRA). In this he urges that the simple, mainly visual, tools he advocates should not be seen just as tools of analysis for visiting professionals. Rather he urges that their use be planned, managed and evaluated jointly with local people (Chambers, 1997) – in our terms, jointly by the project agents and their hosts. Chambers reviews the range of methods used as follows:

- SPATIAL: Mapping & modelling (maps on ground, paper, chalk, pens, symbols)
- NOMINAL: Collecting, naming, listing (collections, cards, symbols, lists)
- TEMPORAL: Sequencing (ground, paper, cards, symbols)
- ORDINAL: Sorting, comparing, ranking (cards, symbols, matrices)
- NUMERICAL: Counting, estimating, comparing, scoring (seeds, stones, matrices)
- RELATIONAL: Linking, relating (Venn diagrams (alternatively known as *Chapati diagrams in India*), cards, symbols, lines).

This list shows that, in the world of participatory development, foundations for the wider use of methods for communicating about the structure of complex problems already exist. These and other participatory methods have now become advocated by such major development sponsors as the World Bank (World Bank, 1996). So an important question for management and systems scientists is whether we can offer any significant extensions to the toolkits of those already involved in the world wide participatory development movement.

ADDED VALUE FROM THE MANAGEMENT AND SYSTEMS SCIENCES?

In a recent Lincoln School of Management working paper (Friend, 1998), I put forward the view that there was a particular opportunity for generating added value in participatory development through what I described as a concept of *reciprocal outreach*. The essence of this concept is expressed in Figure 2 below.

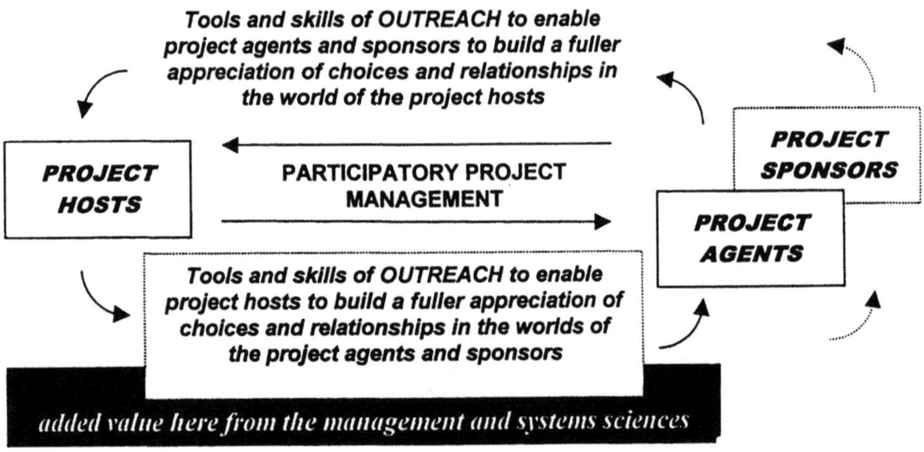

Figure 2. A Principle of Reciprocal Outreach

The initial emphasis behind the toolkit of Rapid Rural Appraisal was on ways of helping agents and sponsors to build a clearer understanding of the complexity of *local* realities – as indicated at the top of Figure 2. This same emphasis has now been carried through into the more participatory toolkit of PRA, with the important difference that project management is now seen as a joint task in which the local hosts should see themselves as active partners.

Yet what about the *reciprocal* challenge of empowering local project hosts in turn to build a clearer understanding of the complex external realities (whether systemic or more transient) of the worlds of the various sponsors and agents of development that may influence the collective choices they might make? This is where my own experience in introducing decision-centred rather than systems-centred communication tools has already demonstrated what can be done. Specifically, the strategic choice approach offers tools with which to identify the most critical sources of *uncertainty* that impinge on current choices – whether these relate to knowledge of external influences, or to the values that should guide decision-making, or to the intended actions of others. If agreement can be reached on the most important external sources of uncertainty, then discussion becomes possible on what kinds of strategy (investigative, political or negotiative) might be adopted in response. Options include anything from informal enquiries by telephone to political lobbying or alliance-building; or indeed major surveys or analyses where these can be justified. A further option, indeed, is the use of more holistic systems methods to explore the internal complexities of any external policy systems the influence of which seems especially critical.

The challenge of designing an agreed outreach strategy may become far from simple in practice. Experience has repeatedly shown that the lives of local people and communities are rarely influenced only by one development project at a time. There may therefore be important influences to appreciate from many other parallel projects, some of them perhaps of a more informal or even covert nature. Each of these may involve a different configuration of agents, sponsors and indeed co-hosts, all with their own complex programme influences and histories to be disentangled, as indicated in Figure 1.

Figure 2 suggests a significant hypothesis relating to participatory development. It is that *one of the most important routes to the empowerment of host communities lies in extending their capacities for OUTREACH towards more understanding of - and influence over - those in agent and sponsor roles that impinge most directly on their local freedom to choose.*

The successes claimed by advocates of participatory development methods have so far been primarily presented in terms of building commitment to communal action at local level. Examples include the formation of local businesses, the building of latrines to reduce public health risks, and steps to enhance the influence of women in community decision-making.

However important such local initiatives may be, a capacity to reach out towards the worlds of external agents and sponsors will become increasingly crucial as the lives of communities become disrupted by new configurations of economic, political or environmental forces. In the developing world – and indeed in many parts of the so-called developed world - the persistence of endemic poverty and disease remains a primary focus for many international development programmes. Yet there are also dynamic forces of *impoverishment,* and disruption of long-standing cultures, that call for a different kind of response. These call for capacities to engage with the complexities of transient problems that exist at this time (*esta*) as well as those that exist (*es*) in a more systemic sense.

The world of international development is already populated by many other professionals, from whom we in OR and systems science have much to learn. They operate in a milieu pervaded by a wider political discourse, relating to the management of an increasingly globalised economy within a highly fragmented political order. In this discourse, such benign concepts as partnership, capacity building, human rights and civil society must contend with more malign forces such as exploitation, environmental degradation, corruption and civil war. In the opening decades of our new millennium, it would be hard to find a more daunting yet exciting challenge to the management and systems sciences.

REFERENCES

Boothroyd, H., 1978, *Articulate Intervention,* Taylor and Francis, London.

Chambers, R., 1983, *Rural Development: Putting the Last First,* Longman Scientific and Technical, Harlow.

Chambers, R., 1997, *Whose Reality Counts? Putting the First Last,* Intermediate Technology Press, London.

Checkland, P.B., 1981, *Systems Thinking, Systems Practice*, Wiley, Chichester.

Faludi, A., 1987, *A Decision-Centred View of Environmental Planning,* Pergamon, Oxford.

Flood, R.L., and Jackson, M.C., 1991, *Creative Problem Solving: Total Systems Intervention*, Wiley, Chichester.

Friend, J.K., 1977, The dynamics of policy change, *Long Range Planning*, 10:40-47.

Friend, J. K., 1998, *Managing Development Projects and Programmes: Fresh Perspectives towards an Action Research Agenda*, Lincoln School of Management Working Paper No. 21, University of Lincolnshire and Humberside, Lincoln and Hull.

Friend, J. K., Bryant, D. T., Cunningham, J. B., and Luckman, J., 1998, Negotiated project engagements: learning from experience, *Human Relations,* 12:1509-1542.

Friend, J. K., and Hickling, A., 1997, *Planning under Pressure: the Strategic Choice Approach* (second edition), Butterworth-Heinemann, Oxford.

Friend. J. K, Power, J. M., and Yewlett, C. J. L., 1974, *Public Planning: the Inter-Corporate Dimension*, Tavistock Publications, London.

Hickling, A, 1989, Gambling with frozen fire? in: *Rational Analysis for a Problematic World* J.V. Rosenhead, ed., Wiley, Chichester.

Lakatos, I., 1972, Falsification and the methodology of scientific research programmes, in: *Criticism and the Growth of Knowledge,* Lakatos, I., and Musgrave, A., eds., Cambridge University Press, Cambridge.

Rosenhead, J. V., ed., 1989, *Rational Analysis for a Problematic World,* Wiley, Chichester.

Stringer, J., 1967, Operational research for multi-organisations, in: *Operational Research Quarterly,* 18:105-20.

World Bank, 1996, *The World Bank Participation Sourcebook*, The World Bank, Washington DC.

MEANING NETWORKS AND VERBAL COMMUNICATION: A SEMIOTIC PERSPECTIVE OF HUMAN DISCOURSE

Marcel Danesi

Program in Semiotics and Communication Theory
University of Toronto
Toronto, Canada.

INTRODUCTION

Verbal communication unfolds so automatically that we hardly ever take notice of the complex conceptual system that makes it possible for us to engage in it so effortlessly. In the late 1940s, the relation between that system and the grammatical system that allows us to transmit information "linearly" in actual speech situations came to constitute a central preoccupation of language and communication scientists. It was the American engineer Claude Shannon (1948) who argued in that era that information of any kind could be described in terms of binary choices between equally probable alternatives. From Shannon's work—and that of mathematician Norbert Wiener (1949), who pioneered the field of cybernetics and the development of computer science—there emerged a widespread notion in the language and communication sciences in the 1950s, still prevalent today, that verbal communication was subject to the same rule-governed stochastic processes that characterize mechanical and animal signaling systems.

By the 1960s, this led to the development of theories of language which continue, to this day, to portray language as a grammatical "object" based on universal stochastic principles. But by the 1980s, and certainly by the 1990s, it became obvious that such an approach to language hardly told the whole story of how grammar delivers concepts in even the most simple speech acts (Emmeche, 1998; Kull, 2000; Brier, 2000). Isolated from speech, grammatical systems can indeed be described in mathematical terms, as these theories have aptly shown. But this tells us nothing about the ways in which the architecture of grammar delivers the complex *meanings* of sentences. Despite substantial and noteworthy research on the nature of grammatical rules and syntactic systems since the publication of Noam Chomsky's *Syntactic Structures* in 1957, current grammatical theories and models seem incapable of ade-

Understanding Complexity, Edited by Ragsdell and Wilby,
Kluwer Academic/Plenum Publishers, New York, 2001

quately explaining the conceptual richness of even the simplest of sentences. This is, no doubt, the reason why Chomsky continues to separate grammatical phenomena from meaning considerations (Chomsky, 2000).

Starting in the 1970s, there emerged several interesting attempts to make grammatical theories more sensitive to the conceptual complexity inherent in speech acts (e.g. Hymes 1971; Halliday, 1975, 1985). This led to much significant work in the 1980s and 1990s examining the relation between conceptual and grammatical structure from a non-stochastic viewpoint (e.g. Fauconnier, 1985, 1997; Langacker, 1987, 1990; Croft, 1991; Ruwet, 1991; Deane, 1992; Taylor, 1995; Fauconnier and Sweetser, 1996; Nuyts and Pederson, 1997; Allwood and Gärdenfors, 1998; Dirven and Verspoor, 1998). The proposal put forward here is in line with such research. Indeed, my objective is to suggest that the categories making up sentence structure in discourse situations are, in effect, "reflexes" of largely unconscious "conceptual networks," whose various "circuits" are converted into specific types of words and phrases in the delivery of speech. The point of departure for such research can be traced to 1977, when Pollio, Barlow, Fine, and Pollio published their extensive investigation of common discourse texts, finding them to be structured primarily by metaphorical concepts. They discovered that speakers of English, for instance, uttered on average 3,000 novel metaphors and 7,000 idioms per week. Their pivotal study was followed by Lakoff and Johnson's momentous 1980 study, *Metaphors We Live By*, which has since provided a powerful framework for relating the meaning of specific grammatical devices to particular conceptual structures (see, for instance, Kövecses, 1986, 1988, 1990; Lakoff, 1987; Johnson, 1987; Sweetser, 1990; Indurkhya, 1992; Danesi, 1993; Gibbs, 1994; Goatley, 1997; Lakoff and Johnson, 1999). To put it in strictly semiotic terms, this line of inquiry has finally started to show how *signifieds*, the units of meaning and reference, can be linked etiologically to their *signifiers*, the physical forms that deliver these units.

Although semioticians have, by and large, been openly critical of the Chomskyan view of language (Sebeok and Danesi, 2000), they have not normally become involved in research designed to provide a viable alternative based on the relation between signifieds and signifiers in language structure. The purpose of the present paper is to offer such an alternative, derived primarily from experimental and pedagogical work with second language learners—individuals who, more than anyone else, are faced with the complex task of learning how the signifiers of the new language overlap with native language signifieds and how they deliver conceptually new and/or subtly different signifieds (Danesi, 2000). The claim to be made here is that verbal discourse involves a "reflexive loop" between the grammatical and the conceptual domains.

SEMIOTIC NETWORK THEORY

Called *langue* by Saussure (1916) and *linguistic competence* by Chomsky (1957), knowledge of language as a rule-governed system has been traditionally assumed to be independent of how it is applied to real-life communicative situations, which Saussure called *parole*. In the early 1970s, the linguist Dell Hymes (1971) challenged this view, proposing that knowledge of language structure was interconnected with knowledge of how to use it appropriately in specific social settings. He called this type of knowledge *communicative competence*. In actual fact, the study of communicative competence was implicit in the work of various structuralist linguists and communication theorists before Hymes (e.g. Firth, 1957;

Jakobson, 1960; Austin, 1962; Dance, 1967; Searle, 1969; Andersch, Staats, and Bostrom, 1969; Barnlund, 1970). It has become a major focus for both sociolinguistics and communication theory ever since (e.g. Myers and Myers, 1985; Cherwitz and Hikins, 1986; Tannen, 1989; Fairclough, 1995; Van Dijk, 1997). At the core of this approach to discourse is the idea of *transaction*—the view that verbal structures in discourse are "negotiated" between the interlocutors and that acts of negotiation influence cumulatively the actual grammar and vocabulary of a language. Such research has shown rather convincingly that communication variables are primary factors in modifying grammar and vocabulary, thus largely rebutting the view that grammatical systems are based on "universal properties" and thus are impervious to the influence of communication. As Colin Cherry aptly put it in his monumental study of communication (1957: 9), the use of language among humans "is essentially a social affair." But, with few exceptions (e.g. Kress, 1985; Ellis and McClintock, 1990; Garza-Cuarón, 1991; Cobley, 1996; Agha, 1997; Kramsch, 1998), lacking from communicative competence theory is the fundamental semiotic view of language as a representational device interconnected with the other (nonverbal) representational systems of a culture (Verschueren, 1995; Yabuuchi, 1996; Edwards, 1997; Danesi and Perron, 1999; Jaworski and Coupland, 2000). This view emphasizes that language is not an autonomous code, separate from the other codes humans employ to represent and communicate information, ideas, emotions, etc. The central notion of what I have elsewhere called *semiotic network theory* (Danesi, 2000) is that there exist three main types of conceptual networks that link language with nonverbal codes—*denotative*, *connotative*, and *metaphorical*. These are reflected in the forms that specific signifiers in each of the codes of a culture assume. In short, the same signifiers surface in the form of different verbal and nonverbal signifiers because these are interconnected to each other by the same complex cognitive circuitry that characterizes the conceptual network systems present in a culture.

Network theory has been developed primarily from several research projects carried out at both the University of Toronto and the University of Lugano during the academic years 1997-1998. Over 500 students were instructed to draw up network analyses of over 200 common concepts, ranging from colors to emotions in English and Italian. Their analyses were then matched against the conceptual structures inherent in common written texts, such as newspaper and magazine articles published in Toronto and Lugano. The findings suggest rather strongly that verbal communication is shaped by a complex web of denotative, connotative, and metaphorical circuits that are concealed in every word, phrase, and sentence (Danesi, 2000).

In Saussurean semiotics, the term *concept* designates the conventional meaning we get from a sign (Saussure, 1916). As it turns out, however, it is not a straightforward matter to explicate what a concept is by *using other words* to do so. Consider, for example, what happens when we look up the definition of a word such as *cat* in a dictionary. Typically, the latter defines a *cat* as "a carnivorous mammal (*Felis catus*) domesticated since early times as a catcher of rats and mice and as a pet and existing in several distinctive breeds and varieties." The problem with this definition is that it uses *mammal* to define *cat*. What is a *mammal*? The dictionary defines *mammal* as "any of various warm-blooded vertebrate *animals* of the class Mammalia." What is an *animal?* The dictionary goes on to define an *animal* as "a living *organism* other than a plant or a bacterium." What is an *organism?* An *organism*, the dictionary stipulates, is "an individual animal or plant having diverse organs and parts that function together as a whole to maintain *life* and its activities." But, then, what is *life? Life*, it specifies, is "the property that distinguishes living *organisms*." At that point it is apparent that the dictionary has gone into a conceptual loop—it has employed an already-used concept, *organism*, to define *life*.

Looping is caused by the fact that dictionaries employ words, which of course encode other concepts, to define an entry. As it turns out, the dictionary approach just described is the only possible one—for the reason that all human systems of knowledge have a looping associative structure, including mathematics, as the brilliant mathematician Kurt Gödel (1906-1978) demonstrated in 1931. This suggests that the meaning of something can only be inferred by relating it to the meaning of something else to which it is, or can be, associated. There simply is no such thing as an "absolute concept." So, the meaning of *cat* is something that can only be extrapolated from the *circuitry* of conceptual associations that it evokes. This circuitry can be called a *network*. In addition to the concepts of *mammal*, *animal*, *organism*, and *life*, used by the dictionary, one can add others such as *whiskers* and *tail* to the circuitry of the *cat* network. In sum, the meaning of a concept such as *cat* crystallizes from an intricate interplay of related conceptual associations that it evokes.

There are several things about networks and network theory that must be made clear from the very outset. First, the term *theory* is not used in its strictly scientific sense, but rather in its original etymological sense of "a view." Network theory is not a "theory of concepts" in the psychological sense. It simply provides a descriptive apparatus for literally *showing* what dictionary makers have known for centuries—namely that the meaning of something is impossible to pin down without reference to other meanings. The position of *nodes* (concepts), the configuration of circuits (the associations among concepts), and the "distances" between nodes and circuits in a network reflect no necessary pattern or intrinsic structure. There is no limit (maximum or minimum) to the number and types of nodes and circuits that can be used to characterize a concept. It depends on a host of factors, not the least of which is the knowledge of the network-maker. In the network for *cat*, secondary circuits generated by *mammal*, for example, could be extended to contain *carnivorous*, *rodent-eater*, etc.; the *life* node could be extended to generate a secondary circuit of its own containing nodes such as *animate, breath, existence*, etc. in no particular order; other nodes such as *feline, carnivorous, Siamese, tabby*, etc. could be inserted to give a more detailed "picture" of the conceptual structure of *cat*; and so on. Finally, network design will vary according to case and necessity. The network described above would put *cat* at its *focal* point because that is the concept under consideration by the dictionary. However, if *animal* were to be needed as the focal concept, then *cat* would be represented differently as a nonfocal node connected to it in a circuit that would also include *dog* and *horse*, among other associated nodes. In effect, there is no way to predict the configuration of a network in advance. It all depends on the analyst, on the purpose of the analysis, on the type of concept, and on other such factors that are variable and/or unpredictable.

The primary node concepts—*mammal*, *animal*, *life*, and *organism*—are superordinate ones; *cat* is a basic concept; and *whiskers* and *tail* are subordinate concepts. In prototype theory (e.g. Rosch 1973), superordinate concepts are those that have a highly general referential function. Basic concepts have a typological function. They allow for reference to types of things. Finally, subordinate concepts have a detailing function. Although it is beyond the purpose of the present discussion, it would be interesting to investigate the relation of nodes and circuits (primary, secondary, etc.) to these functions and determine if a pattern emerges. That is something that will have to be left for future work on semiotic network analysis. Clearly, the configuration of a network will vary according to the function of its focal node—i.e. a network that has a superordinate focal node (e.g. *mammal*) will display a different pattern of circuitry than will one that has a basic concept at its focal center.

TYPES OF NETWORKS

Denotative, connotative, and metaphorical concepts are not to be considered separate phenomena, but rather, interconnected to each other through various kinds of circuitry and network linkages. *Denotation* is the initial meaning captured by a concept. The denotative meaning of the word *blue* in English, for instance, encodes the image of a hue on the color spectrum with a wavelength of approximately 450 to 490 nanometers. The specific image that comes to mind will be different from individual to individual. But all images will fall within the above wavelength, if one is a native speaker of English. The denotative concept of "blueness" is forged cognitively from the experience of observing the hues found in natural phenomena such as the sky and the sea, by observing other hues in things, and so on.

The denotative network for this focal node will thus contain circuits made up of nonfocal nodes such as *color, shade, hue, gradation, sky,* and *sea*, among others. Since *blue* is a type of color, it is really part of a conceptual, or network, domain that has *color* as its focal point. However, in specific network analyses, it is not necessary to show the relevant network domain—in which *blue* would, in effect, be configured as a primary node connected to *color.* A *network domain* can be defined as the associative configuration generated by superordinate categories—*color, animals,* etc. Within such domains, basic and subordinate concepts can be subdivided, for the purpose of a specific analysis, into smaller networks of their own. That applies to the network designed for *blue* as a type of color interconnected to *yellow*, *green*, etc. within the same circuit.

Denotative networks allow speakers of a language to talk and think about concrete things in specific ways. But such networks are rather limited when it comes to serving the need of describing abstractions, emotions, morals, etc. For this reason they are extended considerably through further associative thinking. Consider the use of *cat* and *blue* in sentences such as the following ones:

(1) He's a real cool *cat.*
(2) Today I've got the *blues*.
(3) She let the *cat* out of the bag.
(4) That hit me right out of the *blue*.

These encode *connotative* and *metaphorical* meanings, which are "added" or "extended" meanings of the two concepts. The use of *cat* in (1) to mean "attractive," "engaging," etc. comes out of the network domain associated with jazz *music*; and the use of *blues* in (2) to mean "sad," "gloomy," etc. comes out of the network domain associated with blues *music*. In effect, these have been linked to the original networks of *cat* and *blue* through the channel of specific cultural traditions. They are nodes that interconnect *cat* and *blue* to the network domains of *jazz* and *blues* music. The meaning of "something secret" associated with *cat* in example (3) above and the meaning of "unexpectedness" associated with *blue* in (4) have resulted from linking *cat* with the *secrecy* network domain and *blue* with the *sky* domain. Sentence (3) is, in effect, a specific instantiation of the conceptual metaphor [animals reflect human life and activities], which underlies common expressions such as:

(5) It's a *dog's* life.
(6) Your life is a *cat's* cradle.
(7) I heard it from the *horse's* mouth.

Sentence (4) is an instantiation of the conceptual metaphor [Nature is a portent of destiny]—which literary critics classify as a stylistic technique under the rubric of *pathetic fallacy*. This concept underlies such common expressions as:

(8) I heard it from an *angry wind*.
(9) *Cruel clouds* are gathering over your life.

The networks that are generated by metaphorical signifieds extend the meanings of signs within networks considerably. Comprehensive network analyses of *cat* and *blue* would have to show how all meanings—denotative, connotative, metaphorical—are interconnected to each other through complex circuitry. It is the ability to navigate through the overarching circuitry of such networks, choosing appropriate denotative, connotative, or metaphorical nodes according to communicative need, and integrating them cohesively into appropriate individually-fashioned circuitry to match the need, that constitutes what may be called *conceptual competence* in a language, as opposed to abstract *linguistic competence*.

The connotative extensional process is, needless to say, highly *associative*. But it is not one based on *association-by-sense*, as it is in the formation of denotative concepts. Rather, it is based on *association-by-inference*. To grasp what this means, consider the word *tail*, which the dictionary defines as "the flexible appendage found at the rear end of an animal's body." This is the denotative meaning of *tail* in utterances such as the following:

(10) My cat's *tail* is over one foot long.
(11) Are there any species of dogs without *tails*?
(12) That horse's *tail* is rather short, isn't it?

In a denotative network *tail*, as a focal node, would be connected to a circuit that contains *appendage* and *rear-end* nodes. These provide basic information about what a *tail* is—an extremity —and where it is found on an animal—on its rear end. Now, these nodes are what guide the extension of *tail* to encompass meanings such as following:

(13) The *tail* of that shirt is not bleached.
(14) Do you want heads or *tails* for this coin toss?
(15) The *tail* section of that airplane is making a funny noise.

Such extensions are hardly random or disconnected to the original circuit. Shirts, coins, and airplanes are conceptualized in English-speaking cultures as having appendages and rear ends. In network terms, a *shirt*, a *coin*, and an *airplane* are concepts that belong to separate networks of their own. However, through associative inference these are interlinked to the *tail* network. The process of network linking can be called *grafting*. Grafting is the process that underlies connotation and metaphorization.

As another practical example, consider the following metaphorical statement: "The professor is a *snake*." Clearly, it is not the denotative meaning of the vehicle, *snake*, that is transferred to the topic, *professor*, but rather its connotative meanings, namely the culture-specific characteristics perceived in snakes—"slyness," "danger," "slipperiness," etc. It is this circuit of connotations linked to *snake* that are grafted onto the *professor* circuit. The grafting of the connotative nodes associated with the source network domain circuit (*snake*) onto the focal target domain node (*professor*) is what creates the meaning (or ground) of the metaphor. The concept of grafting suggests that this statement is hardly an isolated example of metaphorical fancy; rather, it implies that it is one of an infinitude of similar expressions that cluster around the idea that [human personality] is understandable in terms of [animal features]:

(16) John is a *pig*.
(17) That woman is a *tiger*.
(18) My friend is a *gorilla*.
(19) She *roars* when she gets angry.

Each is a specific instantiation of that very idea—namely [human personality is understandable in terms of animal features], or simply [people are animals]. This is dubbed a *conceptual metaphor* by Lakoff and Johnson (1980) and, more recently, a *metaform* by Sebeok and Danesi (2000). Note again that the grafting of meanings in the metaphorization process is not based on linking denotative circuits, but connotative ones. Thus, it is not the reptilian physical qualities of snakes, or the feline qualities of tigers, that are grafted onto [people], but rather the kinds of behavioral characteristics that snakes and tigers are thought to have in human terms. This is what creates the meaningful circuitry in metaphorization. It is not a simple transferal process, but one based on *association-by-inference*, as it has been called above. Using electric current as an analogy, it can be said that such circuits run on "alternating conceptual current," so to speak.

Given the controversy surrounding the term *association* in psychology and linguistics, it is necessary to clarify, albeit briefly, what is meant by it in the framework of network theory. In psychology, *associationism* is the theory that the mind comes to know concepts by combining simple, irreducible elements through mental connection. As is well known, interest in associationism was kindled in antiquity by Aristotle (384-322 BC), who recognized four strategies by which associations are forged: through *similarity* (e.g. an orange and a lemon), through *difference* (e.g. hot and cold), through *contiguity* in time (e.g. sunrise and a rooster's crow), and through *contiguity* in space (e.g. a cup and saucer). British empiricist philosophers John Locke (1632-1704) and David Hume (1711-1776) saw sensory perception as the underlying factor in such processes. In the nineteenth century, the Aristotelian view was examined empirically, leading eventually to the foundation of an associationist school of psychology, guided by the principles enunciated by James Mill (1773-1836) in his *Analysis of the Phenomena of the Human Mind* (1829). In addition to Aristotle's original four strategies, that school found that such factors as *intensity*, *inseparability*, and *repetition* added to the strength of an association: e.g. *arms* are associated with *bodies* because they are inseparable from them; *rainbows* are associated with *rain* because of repeated observations of the two as co-occurring phenomena; etc.

The one who developed associationism experimentally was Edward Thorndike (1874-1949), who extended the work initiated by the Russian psychologist Ivan Pavlov (1849-1936) in 1904. Pavlov provided an empirical basis for investigating how associations through repetition are made. When he presented a meat stimulus to a hungry dog, for instance, the animal would salivate spontaneously, as expected. This was termed the dog's "unconditioned response." After Pavlov rang a bell while presenting the meat stimulus a number of times, he found that the dog would eventually salivate only to the ringing bell, without the meat stimulus. Clearly, Pavlov suggested, the ringing by itself, which would not have triggered the salivation initially, had brought about a "conditioned response" in the dog. By *association* the dog had learned something new. Every major behavioral psychologist has utilized the Pavlovian notion of associationism in one way or other. Although behaviorists believe all thought processes can be accounted for through associations of stimuli and responses, other psychologists strongly reject such an approach as inadequate to explain creative thought and verbal behavior.

The meaning of *association* as used in the network theory framework is not the Pavlovian one. In line with twentieth century Gestalt psychology, it is used to stress that abstract concepts beget their meanings only in relation to other concepts. Gestalt psychologists believed that pattern, or form, was the most important part of experience. The whole pattern in a conceptual network, for instance, gives meaning to each individual element (node, circuit, etc.) within it. In other words, the whole is more important than the sum of its parts. As discussed above, network patterns can be forged by sense, i.e. by observing physical features of referents, or by inference, i.e. by applying the sense associations to referents that are perceived as possessing the same features.

REFLEXIVIZATION

Associative conceptual structure is converted into linear surface grammatical structure through a process that can be called *reflexivization*. Consider, for example, an underlying circuit containing *snake* as a metaphorical concept that is to be used in a specific speech act. In the surface language that is chosen to deliver it, it can show up as a verb (20), if it is the snake's movements that are grafted onto the target, or as an adjective (21), if it is a serpentine quality that is grafted conceptually onto the target:

(20) The professor *snaked* his way around the issue.
(21) The professor has a *snaky* way of doing things.

The difference between the two surface forms—*snaked* and *snaky*—can be traced to underlying circuits that extend the *snake* concept in specific ways. The reflexivization of (20) shows that the grafted concept included a *movement* node. In (22) the relevant circuit grafted onto the target concept included, instead, a *quality* node.

The notion of reflexivization is not a theory of grammar. It is a heuristic technique for showing how words, phrases, and sentences appear to *reflect* conceptual structure, i.e. to encode it in specific ways. Needless to say, surface linear structure reflects not only concepts, but is also sensitive to communicative functions, situational variables, stylistic needs, etc. There are an infinitude of ways in which the reflexivization of concepts can unfold. The choices made by the speaker, the context of the speech act, the grammatical and lexical knowledge of the speaker, etc. are the factors that constrain surface structure outcomes. It is not the purpose of reflexivization analysis to consider these factors. The main objective of such analysis is showing how grammar, vocabulary, and concepts are interconnected in a systemic way.

Differences in surface linear structure are typically due to underlying conceptual dichotomies. In Italian, for instance, the difference between the denotative and connotative meaning of an adjectival concept is sometimes reflected in the surface by a difference in position with respect to the noun. Thus, in (22) and (23), the different surface position of the adjective is a reflex of the different networks to which the meanings belong:

(22) Lui è un uomo *povero* ("He's an indigent man").
(23) Lui è un *povero* uomo ("He's a forlorn man").

In (22) it is the denotative meaning of *povero* that is reflected in the surface by a post-positioning of the adjective with respect to the noun (the normal position for qualitative adjectives). In (23) the connotative meaning of *povero* is indicated by means of its pre-positioning, alerting the interlocutor in an anticipatory fashion as to the type of concept that is intended.

As another example of how conceptual dichotomies are reflexivized, consider the use of the English prepositions *since* and *for* in sentences such as the following:

(24) I have been living here *since* 1980.
(25) I have known Lucy *since* November.
(26) I have not been able to sleep *since* Monday.
(27) I have been living here *for* twenty years.
(28) I have known Lucy *for* nine months.
(29) I have not been able to sleep *for* seven days.

An analysis of the complements that follow *since* or *for* reveals that those that follow the former are [points in time], i.e. they are complements that reflect a conception of time as a [point] on a [timeline] which shows specific years, months, etc.: *1980*, *November*, *Monday*, etc. Complements that follow *for*, on the other hand, reflect a conception of time as a [quantity]: *twenty years*, *nine months*, *seven days*, etc. These two network domains—[time is a point] and [time is a quantity]—have an underlying metaphorical circuitry structure, reflecting our propensity to imagine an abstract notion such as "time" in terms of something concrete. These can now be seen to have a specific effect at the level of syntax by motivating a grammatical dichotomy—complements introduced by *since* are reflexes of the conceptual domain [time is a point]; those introduced by *for* are reflexes of the conceptual domain [time is a quantity]. This is, in fact, the kind of *rule of grammar* that reveals how concepts are encoded linearly—it relates, in effect, how two specific domains of conceptualization have worked their way into the grammar. In a word, this rule stipulates how a grammatical dichotomy *reflects* a conceptual dichotomy.

Take, as one final example, the selection of certain verbs in particular types of sentences in Italian. The verb *fare* "to make" is used to convey a weather situation—*fa caldo* (literally) "it makes hot," *fa freddo* (literally) "it makes cold." The physical states of [hotness] and [coldness] are conveyed instead by the verb *essere* "to be" when referring to objects—*è caldo* "it is hot," *è freddo* "it is cold"—and by *avere* "to have" when referring to people—*ha caldo* "he/she is hot," *ha freddo* "he/she is cold." The use of one verb or the other—*fare, essere,* or *avere*—is motivated by an underlying metaphorical conceptualization of bodies and the environment as [containers]. So, the [containment context] in which the quality of [hotness] or [coldness] is located determines the verbal category to be employed. If it is in the environment, it is "made" by Nature *(fa caldo/freddo)*; if it is in a human being, then the body "has" it *(ha caldo/freddo)*; and if it is in an object, then the object "is" its container *(è caldo/freddo)*.

To summarize, it is obvious that conceptual domains leave their reflexes in the grammars of specific languages. Knowledge of such differentiated reflexive properties is what guides conceptually-appropriate communication among interlocutors. *Grammar* in this framework is definable, therefore, as a system that reflexivizes conceptual circuitry in specific ways. It must be emphasized again that the notion of reflexivization is not a theory of grammar. It is proposed simply to show that in the same way that a painting is much more than an assemblage of lines, shapes, colors, and melodies a combination of notes and harmonies, so too a sentence in language is much more than an assemblage of words and phrases built from some abstract rule system in the brain. We use the surface grammatical and lexical codes at our disposal to model the world of concepts in ways that parallel how musicians use melodic elements and painters visual ones to model it.

DISCOURSE

Network analyses of conversations show, above all else, that discourse is structured largely by connotative inter-network linkages. A truly interesting feature of discourse that this type analysis also reveals can be called *connotative chaining*. This occurs when a specific connotative node generates derivative associations in the immediate span of the discourse. To put it figuratively, it tends to "infect" the whole conversation. For instance, once a word such as *drop* is used connotatively by a speaker in a certain situation, then it may spawn a chain of associated concepts such as *pick up, let go*, etc. Here is an example of a conversation excerpt that I recorded at the University of Toronto, during which a speaker (a university student) used the word *drop* as just described: "Yeah, I *dropped* that course yesterday...No, I won't *pick* it *up* next year...The main reason for *letting it go* was the prof. He was awful...Believe me, I haven't *lost* anything..."

In this sample of discourse, the connotative meaning of *drop* initiated a circuit on its own that included *pick up*, *let go*, and *lose* in close proximity to each other. In effect, the image of "falling" is distributed in the circuit, surfacing in various lexical forms. The nodes are linked again through a process of *association-by-inference—picking up* something means that it was *dropped*; *losing* something elicits the image of *dropping* it; and, of course, *letting* something *go* will cause it to *drop*. The construction of the circuit is a subjective act, based on grafting nodes from network domains. This is what makes discourse unpredictable in actual form, but understandable, and even predictable, conceptually. Once connotative circuits have been introduced into discourse they tend to guide the flow of conversation through chaining. In the above circuit, for instance, the *pick up* node led a little later in the conversation to the use of *take*, which, in turn, generated its own circuitry with two nodes—*carry* and *heavy*: "I really can't *take* any more subjects...I'm already *carrying* the maximum...I've got quite a *heavy* load..."

There are various kinds of connotative chains that characterize discourse flow. Some of these contain nodes based on narrative traditions; these are concepts referring to themes, plot-lines, characters, and settings that surface in narratives. Calling someone a *Casanova* or a *Don Juan*, rather than *lady-killer*, evokes an array of socially-significant connotations that these characters embody. Referring to a place as *Eden* or *Hell* elicits connotations that have a basis in mythic and religious narrative. The circuits that have been grafted from these stories also surface constantly in common discourse events. Climatologists, for example, refer to the warming of the ocean surface off the western coast of South America that occurs every 4 to 12 years when upwelling of cold, nutrient-rich water does not occur as a person, *El Niño,* "the little one" in Spanish. This mythic personification of a climatological condition makes it much more understandable in human terms. Although people do not think of *El Niño* as a mythic figure, they nonetheless find it convenient to blame "him" for certain weather repercussions as if it were one. This is how original myths worked cognitively—the difference being that the personified conditions of the past were actually believed to be real gods or mythical beings. The discourse that surrounds *El Niño* is virtually always interpretable in mythic terms. For instance, I recorded a weather commentary on American television recently that contained a circuit generated by mythic personification: "This year *El Niño* is having a great time of it. *He* has *wreaked havoc upon* anyone or anything in *his* path. *He* has *come down* very strenuously upon us." In effect, connotative and metaphorical circuitry in discourse is densely distributed through networks present in the entire system of culture. Take, as an example, the *up-down* metaphorical concept that entails the connotative feature [verticality]. In verbal discourse this feature is a node that is reflexivized commonly in expressions such as the following:

(30) I'm feeling *up*.
(31) They're feeling *down*
(32) I'm working my way *up* the ladder of success
(33) His enthusiasm has gone *down* considerably.

This same concept manifests itself in the religious domain, where goodness, spirituality, and heaven are portrayed as *up*, and evil, damnation, and hell as *down* in sermons, theological narratives, religious visual representations, the design of churches, etc. In public building design, too, it can be discerned in the fact that the taller office buildings in a modern city are the ones that indicate which institutions (and individuals) hold social and economic power. In musical composition, higher tones are typically employed to convey a sensation of happiness, lower ones of sadness. During speech, the raising of the hand designates notions of amelioration, betterment, growth, etc., whereas the lowering of the hand designates the opposite notions. In bodily representation and perception, this concept shows up in the common viewpoint that *taller is more attractive/shorter is less attractive.* In mathematical and scientific representational practices its reflexivization can be seen, for instance, in the ways in which graphs are designed—lines that are oriented in an upward direction indicate a growth or an increase of some kind, while those that are slanted in a downward direction indicate a decline or decrease.

The foregoing analysis of interconnected networks is intended to show how highly complementary abstractions are to each other and how they are utilized to generate representational practices and systems. The [people are animals] network discussed above is the source of such symbolic activities as the use of animals in totemic codes, in heraldic traditions, in the creation of fictional characters for use in story-telling to children *(Bugs Bunny, Foghorn Leghorn, Daffy Duck*, etc.), in the naming of sports teams (Chicago *Bears*, St. Louis *Cardinals*, Miami *Dolphins*, etc.), and in the creation of surnames, to mention but a few.

This type of analysis also explains why discourse texts produced by computers and foreign-language learners alike (at the beginning of their study of a new language) will manifest a high degree of accuracy in sentence-formation, but they will invariably lack the conceptual appropriateness that characterizes the corresponding discourse texts of native speakers. To put it another way, students and machines "speak" artificially with the formal grammatical structures of the language as they have been taught them or programmed to do respectively, but they are unable to "think" in terms of the conceptual system that underlies the structures: i.e. students typically make-up target language sentences as artificial "carriers" of their own native language concepts through the rules they have been taught; computers generate them in response to the rules programmed into them. When these coincide with the ways in which concepts are relayed by native speakers *naturally*, then the student and machine texts coincide serendipitously with culturally-appropriate discourse texts; when they do not, they manifest an asymmetry between sentence form and conceptual content.

CONCLUDING REMARKS

The notions of semiotic networks and of reflexivization raise some specific questions for future research. The guiding question is: What are the verbal cues that reveal conceptual domains? In this paper, for instance, the structures *since* and *for* were related to the conceptual domain as reflexes of differentiated metaphorical networks: [time is a point] and [time is a quantity]. The work on cognitive grammar by Langacker (1987, 1990), as mentioned, is leading the way in showing us how analyses of this type might be envisioned. Another question is to determine to what extent and in what ways conceptual networks relate to, or are embedded in, world knowledge. Is world knowledge built up from such circuitry? And if so, how is this incorporable into an extensive analysis of language? Some possibilities have been explored in the past (e.g. Pike 1967), and I believe that this kind of exploration is the wave of the future in semiotics, linguistics, and communication science. As Levin (1988:10) has aptly remarked, however, one must proceed cautiously in this area of inquiry, simply because the many modes of knowing defy the possibility of envisioning a single theory—e.g. innate knowledge, personal knowledge, tacit knowledge, spiritual knowledge, declarative and procedural knowledge, knowing that and knowing how, certitude (as well as certainty), and so on. The more appropriate goal for linguistics and semiotics should be, therefore, to determine to what extent and in what specific ways language reflects knowledge structures.

The bulk of the work on grammatical systems in linguistics and verbal communication generally has traditionally excluded the relation between concepts and grammatical categories. The present study has aimed to show, however, that sentence form is shaped by conceptual factors much more than traditional grammatical analysis would allow. The effect of conceptual structure on categorization in grammar has been taken up somewhat in the linguistic literature, but it has never really penetrated the mindset of language scientists until recently. The philosopher Herder (1744-1803), for instance, saw an intimate connection between language and what he called "ethnic character." Subsequently, Wilhelm von Humboldt (1767-1835) gave Herder's hypothesis a more testable formulation when he portrayed the structure of a particular language as interdependent with the thought and behavior of the people using it for communication. Needless to say, von Humboldt's perspective went contrary to the views of the Port-Royale grammarians who saw language as the product of the universal logical laws of the human mind. It was shortly after von Humboldt's alluring pronouncements that the study of "language and mind" was given its first scientific research impetus. In the twentieth century, work on language and thought was pursued by both those espousing a Humboldtian perspective—Sapir, Whorf, and others—and those advocating a universalist Port Royale perspective—especially Chomsky. As mentioned, the goal of Humboldtians, such as Sapir and Whorf, has never been truly envisioned by mainstream linguistics until very recently. The North American version of linguistic science took its characteristic shape and methodological orientation from Leonard Bloomfield's 1933 textbook entitled simply *Language.* In the same way that Euclid's *Elements* bestowed systematicity and unity upon the study and practice of geometry in antiquity through its coherent synthesis of geometrical concepts and techniques, so too did *Language* provide the fledgling science of linguistics in the 1930s with an organized repertory of notions and procedures for carrying out detailed investigations and analytical characterizations of specific languages. This is the main reason why, in my view, Bloomfield's *Language*, and not the work bearing the same title and published more than a decade earlier by Edward Sapir (1921), came to be accepted by the majority of linguists as the point of reference for conducting empirical research and for developing models of language design.

While Bloomfield's work constituted the first true "textbook" in the history of linguistic science, Sapir's book was the first real attempt to provide a framework for studying the relation of language to cognition and culture. And whereas for most of the twentieth century linguists diligently pursued the investigation of language systems *per se*, along the lines laid down first by Bloomfield and later by Chomsky, they have recently started to move more and more toward the adoption of Sapir's original paradigm.

The question that Sapir sought to answer throughout his life is probably as old as civilization itself: How is thought related to language? He was intrigued, in other words, by the possibility that human ideas, concepts, feelings and characteristic social behaviors might be mirrored by the verbal categories that specific cultures employ to codify them. Sapir suspected that the most direct route to the mind was through language. Due to his tragically early death, Sapir was never able to design and carry out a research program aimed at examining his idea rigorously and systematically. As is well known, it fell on the shoulders of Sapir's brilliant student Benjamin Lee Whorf (1956) to elaborate substantively upon his mentor's views and to give them a more empirically-testable articulation. Whorf posited, in essence, that the categories of one's particular language are much more than simple mediators of thought. He saw them as being the "shapers" of the very thought patterns they embody: "The world is presented in a kaleidoscopic flux of impressions which has to be organized by our minds—and this means largely by the linguistic systems in our minds" (Whorf 1956: 153). But Whorf's experimental program for studying the language-thought nexus could not have been devised without his teacher's profound insights. Sapir saw language as being much more than a communication system. He considered it to be a kind of cognitive filter through which humans come to perceive and understand the world around them.

Semiotic network theory is an attempt to make good on Sapir's agenda for linguistic science. It is just one way that can be envisaged for relating formally how thought and language are interdependent. Grammatical properties cannot be studied in isolation. To rewrite natural grammars with reflexive rules would imply research on the domains of meaning that are implicit in sentences first and then to connect the grammatical categories to these in direct ways. Some of the ways have been discussed tentatively in this paper. There is no doubt that future work in semiotics and linguistics will show how to encompass all the traditional morphological and syntactic categories within a semiotic framework of meaning.

Within this framework, everything form verb tenses to adverb usage will be linked to the interconnected experiences of the world that are manifested in the use of a language by native speakers in cultural contexts. This was Sapir's research challenge; and the time has come to take up his challenge seriously.

REFERENCES

Agha, A., 1997, Concept and communication in evolutionary terms, *Semiotica* 116: 189-215.

Allwood, J., and Gärdenfors, P., eds., 1998, *Cognitive Semantics: Meaning and Cognition*, John Benjamins, Amsterdam.

Andersch, E. G., Staats, L. C., and Bostrom, R. C., 1969, *Communication in Everyday Use*, Holt, Rinehart & Winston, New York.

Austin, J. L., 1962, *How to Do Things with Words*, Harvard University Press, Cambridge, MA.

Barnlund, D. C., 1970, A transactional model of communication, in: *Foundations of Communication Theory*, K. K. Sereno and C. D. Mortensen, eds., Harper & Row, New York.

Bloomfield, L., 1933, *Language*, Holt, New York.

Brier, S., 2000, Biosemiotics as a possible bridge between embodiment in cognitive semantics and the motivation concept of animal cognition in ethology, *Cybernetics & Human Knowing* 7: 57-76.

Cherry, C., 1957, *On Human Communication*, MIT Press, Cambridge, MA.

Cherwitz, R., and Hikins, J., 1986, *Communication and Knowledge: An Investigation in Rhetorical Epistemology*, University of South Carolina Press, Columbia.

Chomsky, N., 1957, *Syntactic Structures,* Mouton, The Hague.

Chomsky, N., 2000, *New Horizons in the Study of Language and Mind,* Cambridge University Press, Cambridge.

Cobley, P., ed., 1996, *The Communication Theory Reader*, Routledge, London.

Croft, W., 1991, *Syntactic Categories and Grammatical Relations*, University of Chicago Press, Chicago.

Dance, F., 1967, *Human Communication Theory*, Holt, Rinehart & Winston, New York.

Danesi, M., 1993, *Vico, Metaphor, and the Origin of Language,* Indiana University Press, Bloomington.

Danesi, M., 2000, *Semiotics in Language Education*, Mouton de Gruyter, Berlin.

Danesi, M., and Perron, P., 1999, *Analyzing Cultures: A Handbook and Reference Manual*, Indiana University Press, Bloomington.

Deane, P., 1992, *Grammar in Mind and Brain: Explorations in Cognitive Syntax*, Mouton de Gruyter, Berlin.

Dirven, R., and Verspoor, M., 1998, *Cognitive Exploration of Language and Linguistics*, John Benjamins, Amsterdam.

Edwards, D., 1997, *Discourse and Cognition*, Sage, London.

Ellis, R., and McClintock, A., 1990, *If You Take My Meaning: Theory into Practice in Human Communication*, Arnold, London.

Emmeche, C., 1998, Defining life as a semiotic phenomenon, *Cybernetics & Human Knowing* 5: 33-43.

Fairclough, N., 1995, *Critical Discourse Analysis: The Critical Study of Language*, Longman, London.

Fauconnier, G., 1985, *Mental Spaces*, Cambridge University Press, Cambridge.

Fauconnier, G., 1997, *Mappings in Thought and Language*, Cambridge Univ. Press, Cambridge.

Fauconnier, G., and Sweetser, E., eds., 1996, *Spaces, Worlds, and Grammar*, University of Chicago Press, Chicago.

Firth, J. R., 1957, *Papers in Linguistics: 1934-1951*, Oxford University Press, Oxford.

Garza-Cuarón, B., 1991, *Connotation and Meaning*, Mouton de Gruyter, Berlin.

Gibbs, R. W., 1994, *The Poetics of Mind: Figurative Thought, Language, and Understanding*, Cambridge University Press, Cambridge.

Goatley, A., 1997, *The Language of Metaphors*, Routledge, London.

Gödel, K., 1931, Über formal unentscheidbare Sätze der Principia Mathematica und verwandter Systeme, Teil I, *Monatshefte für Mathematik und Physik* 38: 173-189.

Halliday, M. A. K., 1975, *Learning How to Mean: Explorations in the Development of Language*, Arnold, London.

Halliday, M. A. K., 1985, *Introduction to Functional Grammar*, Arnold, London.

Hymes, D., 1971, *On Communicative Competence*, Univ. of Pennsylvania Press, Philadelphia.

Indurkhya, B., 1992, *Metaphor and Cognition*, Kluwer, Dordrecht.

Jakobson, R., 1960, Linguistics and poetics, in: *Style and Language*, Thomas A. Sebeok, ed., MIT Press, Cambridge, MA.

Jaworski, A., and Coupland, N., eds., 2000, *The Discourse Reader*, Routledge, London.

Johnson, M., 1987, *The Body in the Mind: The Bodily Basis of Meaning, Imagination and Reason*, University of Chicago Press, Chicago.

Kövecses, Z., 1986, *Metaphors of Anger, Pride, and Love: A Lexical Approach to the Structure of Concepts*, John Benjamins, Amsterdam.

Kövecses, Z., 1988, *The Language of Love: The Semantics of Passion in Conversational English*, Associated University Presses, London.

Kövecses, Z., 1990, *Emotion Concepts*, Springer, New York.

Kramsch, C., 1998, *Language and Culture*, Oxford University Press, Oxford.

Kress, G., 1985, *Linguistic Processes in Sociocultural Practice*, Deakin Univ. Press, Melbourne.

Kull, K., 2000, Organisms can be proud to have been their own designers, *Cybernetics & Human Knowing* 7: 45-56.

Lakoff, G., 1987, *Women, Fire, and Dangerous Things: What Categories Reveal about the Mind*, University of Chicago Press, Chicago.

Lakoff, G., and Johnson, M., 1980, *Metaphors We Live By*, Chicago University Press, Chicago.

Lakoff, G., and Johnson, M., 1999, *Philosophy in the Flesh: The Embodied Mind and Its Challenge to Western Thought*, Basic, New York.

Langacker, R. W., 1987, *Foundations of Cognitive Grammar*, Stanford Univ. Press, Stanford.

Langacker, R. W., 1990, *Concept, Image, and Symbol: The Cognitive Basis of Grammar*, Mouton de Gruyter, Berlin.

Levin, S. R., 1988, *Metaphoric Worlds*, Yale University Press, New Haven.

Myers, G. E., and Myers, M. T., 1985, *The Dynamics of Human Communication*, McGraw-Hill, New York.

Nuyts, J., and Pederson, E., eds., 1997, *Language and Conceptualization*, Cambridge University Press, Cambridge.

Pike, K., 1967, *Language in Relation to a Unified Theory of the Structure of Human Behavior*, Mouton, The Hague.

Pollio, H., Barlow, J., Fine, H., and Pollio, M., 1977, *The Poetics of Growth: Figurative Language in Psychology, Psychotherapy, and Education*, Lawrence Erlbaum Associates, Hillsdale, N.J.

Rosch, E., 1973, On the internal structure of perceptual and semantic categories, in: *Cognitive Development and Acquisition of Language*, T. E. Moore, ed., Academic, New York.

Ruwet, N., 1991, *Syntax and Human Experience*, University of Chicago Press, Chicago.

Saussure, F. de, 1916, *Cours de linguistique générale*, Payot, Paris.

Searle, J. R., 1969, *Speech Acts: An Essay in the Philosophy of Language*, Cambridge University Press, Cambridge.

Sebeok, T. A., and Danesi, M., 2000, *The Forms of Meaning: Modeling Systems Theory and Semiotic Analysis*, Mouton de Gruyter, Berlin.

Shannon, C., 1948, A mathematical theory of communication, *Bell Systems Technical Journal* 27: 379-423.

Sweetser, E., 1990, *From Etymology to Pragmatics: The Mind-as-Body Metaphor in Semantic Structure and Semantic Change*, Cambridge University Press, Cambridge.

Tannen, D., 1989, *Talking Voices*, Cambridge University Press, Cambridge.

Taylor, J. R., 1995, *Linguistic Categorization: Prototypes in Linguistic Theory*, Oxford University Press, Oxford.

Van Dijk, T., ed., 1997, *Discourse as Social Interaction*, Sage, London.

Verschueren, J., 1995, Linguistic pragmatics and semiotics, *Semiotica* 104: 157-170.

Wiener, N., 1949, *Cybernetics, or Control and Communication in the Animal and the Machine*, MIT Press, Cambridge, MA.: Illocution, interaction, or text? *Semiotica* 110: 197-229.

CAN RELIGION BE RECONCILED WITH SCIENCE?

John F. Haught

Georgetown University
Washington, DC 20057-1135

The central issue in science and religion is whether we can any longer plausibly think of the cosmos as purposeful. Instead of using the term "purpose" we could also speak here about the "point" of the universe in order to enter a conversation initiated by Steven Weinberg's (1977) oft-cited remark that the more comprehensible the universe becomes to physics the more "pointless" it also seems to be. Religions, at least for the most part, have taught us that the universe is permeated by a principle of order or "rightness." In biblical circles, for example, the universe is the expression of a divine "Wisdom" or an eternal Logos, and in Eastern religious thought nature is deeply connected to ultimate principles known as *Rta*, *Dharma* or the *Tao*. The cosmos, in these traditional teachings, is more than just blind and meaningless commotion. It is the expression of a transcendent Meaning (Haught, 2000).

According to most religious traditions, in order to have a meaning or purpose the universe must be structured hierarchically. That is, it must consist of at least several levels, typically thought of--moving from lower to higher--as the inanimate, the living, the sentient and the self-aware. Presiding over all of these levels is a hidden but eminently real Source of meaning, identified in theistic faiths as "God." More often than not, there are also intermediary levels of beings, such as angels, occupying the vast expanse between the human and the divine.

The persistent attraction of a hierarchical scheme consists of the fact that it ensconces the temporal world within the framework of an eternal, absolute and sacred reality immune to transiency and death. Only participation in such a Permanence could give meaning to the perishable flux of finite existence. In fact, the term "hierarchy," which today may seem to be redolent of patriarchal politics, originally implied nothing more than that all things have their origin or principle (*arche*) in the domain of the sacred (*hier*). Traditional religion and philosophy almost unanimously resisted the idea that the universe of being could be reduced to a single, desacralized dimension, and so they consistently clung to a hierarchical vision of things.

Understanding Complexity, Edited by Ragsdell and Wilby,
Kluwer Academic/Plenum Publishers, New York, 2001

However, modern science now seems to have made the classical hierarchical vision indefensible, and in doing so it has made it more difficult for many of us to think seriously of nature as the expression of an eternal Meaning. Perhaps it isn't too much of an oversimplification to say that the main threat modern science has posed to religion is its apparent collapsing of the sacred hierarchical representation of being without which it is difficult to imagine how the universe could have any abiding purpose. In the absence of a hierarchical vision, how can there be any sense of the participation of the temporal in the eternal, and therefore any basis for attributing transcendent value, importance or meaning to the physical universe?

Two features of modern science have been especially instrumental in leveling the ancient hierarchical vision. These are the *atomizing* and the *historicizing* of nature. The ancient Democritean atomism, which Plato and Aristotle had pretty much laid to rest for centuries, has now returned to claim victory in the modern scientific picture of the world. The intuition that life and mind will be made fully intelligible only by scientifically specifying their subordinate physical particulars has made many in the scientific community wonder if there is any more to living and thinking beings than just atomic, molecular or genetic activity, and whether atomic "matter" might not be all there is to reality. If so, then there seems to be little, if any, room for God or a religious sense of purpose.

Atomism, by which I mean the ideal of explaining things in terms of some elemental unit--whether physical particles, molecules, cells, genes, memes, or whatever--makes hierarchy seem either illusory or irrelevant. Scientists today, of course, acknowledge that the physical domain is itself, in a manner of speaking, hierarchically structured. But this kind of hierarchy is trivial and incidental in comparison with the traditional metaphysical sense of discrete levels of being reaching upward and participating in a timeless Absolute. Scientists are now more aware than ever that complex physical systems preside over less complex subsystems, and so there is no question about this rather inconsequential kind of hierarchical structuring of nature. I say "inconsequential" because the higher natural systems are generally pictured as "emerging" derivatively from lower ones without the agency of any supervening principle working from above. Even those who speak today of the "science of complexity" have not always completely forsaken the atomistic agenda.

The second, and perhaps more serious, challenge by science to the religious sense of hierarchy--and consequently to cosmic meaning--lies in its relatively recent discovery that life and mind have emerged rather late in the incredibly long *historical* process we now call "evolution." Evolution is an extremely unsettling idea for a number of reasons, but perhaps most of all because it implies that life and mind have "emerged" only gradually and even accidentally from the lifelessness of physical simplicity. And if the allegedly "higher" levels--life and mind--don't appear in evolution until much later than their lifeless constituents, how can they be in any sense causal or explanatory of the earlier phases of natural history, as they were in the traditional religious framework? If life and mind don't yet *exist* at the earlier stages of evolution, how can they be said to endow with meaning or purpose the process that led to their appearance? Aren't they instead subject to explanation themselves in terms of the lifeless, mindless and inherently meaningless molecular components and physical processes out of which they have only very sluggishly emerged?

That higher levels can emerge from lower levels all by themselves, without any additional causal agency, would have been inconceivable to proponents of the classic hierarchy. The idea that lifeless matter could by itself give birth to life would have violated the principle of causation--that no effect can be greater than its cause. Causation was pictured as flowing downward from the highest level, that of the divine, to the lowest levels, and holding the chain of being up from above, as it were, so that first principles remained intact. But all of this was before we knew *how much time* the cosmos has had to produce life and mind. In the absence of any sense of the vast epochs of natural history that seem now to allow for a *gradual* emergence by trial and error experimentation, the most efficient way of accounting for the complexity of higher systems was to suppose the connivance of the supernatural. And even after the birth of modern science it was not until the relatively recent flowering of geology, paleontology and astrophysics that we discovered deep time and began to suspect that even the most improbable occurrences can take place quite godlessly within its unfathomable compass.

In the absence of immense epochs of time, the emergence of life seemed improbable, and so it made sense to posit the exceptional intervention of a special creative agency. But what if the "unlikely" emergence of life has had billions of years available to it--and if we multiply universes, maybe billions more. Today, it seems, time itself has become so unimaginably protracted that in the thinking of some scientists this temporal vastness itself has tacitly become the demiurge that transforms dead "matter" into living and thinking beings without our having to invoke anything miraculous. Time's very magnitude has placed in question the need for a hierarchical cosmology. After Darwin, vitalism had its appeal briefly because it apparently salvaged the ancient principle of causation, positing a supernatural force as the creative source of life. Today, however, vitalism has been largely discredited, and time itself has become enshrined as a sufficiently creative principle in place of the divine.

In abstraction from our current awareness of deep time, living phenomena seemed to require the superintending creativity of an ineffable divine reality. There is a silent assumption, especially among evolutionary scientists today, that without our new awareness of immeasurable spans of cosmic time, life's creativity would seem extraordinary. But if the enormity of time itself can provide ample opportunity for experimentation, then what need is there to appeal any longer to the idea of a timeless transcendent realm that would draw things together from up above? As the philosopher Daniel Dennett has put it, there is no need for any such "skyhooks" to explain what can be constructed mechanistically over the course of time with "cranes" alone, working from below. Skyhooks made sense only as long as we were ignorant of deep time and the generous scope it allows for trial and error in the long march of matter from a lifeless cosmic past. The discovery of time's immensity, however, allows us to horizontalize and finally altogether dissolve the old hierarchical picture.

As a consequence of the atomizing and historicizing of nature, for many intelligent moderns there apparently no longer remains any explanatory authority in the traditional hierarchical vision with its function of assigning to each level of being a special kind of value, and of endowing the entire cosmic edifice with eternal meaning. The *arche* in the term "hierarchy" used to imply not so much a chronological beginning as a sacred *principle* of explanation. Modern science, however, has looked not to a Platonic realm up above but to the dead mechanical past for the *arche* of all things, including those we value most. In horizontalizing what used to be thought of as an essentially vertical arrangement of being and value, it has challenged the cultural, ethical and religious framework on which human life on this planet had been based for ages.

We cannot exaggerate the dramatic enormity of this great transformation in our human understanding of nature. It seems to me, then, that the central task of religious thought in this age of evolutionary science and cosmology is to face as honestly as it can the question of whether the hierarchical vision, which formed the backbone of our religious traditions and allowed for a purposeful universe, is in any logically coherent sense recoverable today. Can we, without slipping into logical contradictions or denying the clear results of natural science, still clearheadedly profess that the universe is the expression of an Absolute Meaning?

In a way, the question of the reconcilability of religion with science comes down to whether it is both logically and theologically possible for us to embrace simultaneously the hierarchical sense of being on the one hand, and the atomizing and historicizing scientific viewpoint on the other? Can we completely assent to the horizontal understanding of life and mind as emerging chronologically out of "matter," without at least implicitly surrendering our traditional metaphysical sense of their sacred origin. A hierarchical view of reality seems essential to any satisfactory conception of cosmic meaning. A way of thinking about the world in which subordinate dimensions are quietly informed by more comprehensively real and valuable ones is fundamental to any religious intuition that life and mind have a sacred origin or that they are more intrinsically significant than the inanimate stuff they are made of. To deny the reality of a sacred *Arche* or "Principle," would make it very difficult for us to attribute enduring importance to anything. But can we reconcile the hierarchical with the scientific view of the world?

THE RELEVANCE OF INFORMATION

In dealing with this troublesome question I have found considerable help in some ideas of the philosophers Michael Polanyi and Alfred North Whitehead. Here I can provide only a brief summary of their important reflections.

I would begin by noting once again that a kind of hierarchical understanding of nature is not completely absent today even within the world of science. Today even some of the most obstinate reductionists acknowledge that "lower" levels of physical reality are nested within "higher" ones. Molecules enfold atoms; chains of DNA in living cells pattern amino acids; organisms incorporate cells; ecosystems contextualize individual organisms, etc. Reference to the complex hierarchical arrangement of nature is now becoming part of standard scientific discourse. The spirit of reductionism, of course, is still very much alive, but it has been subdued somewhat by a growing awareness of the practical difficulty, if not impossibility, of accounting for higher levels completely or solely in terms of the lower. There is a suspicion that more complex levels cannot simply be understood in terms of the less complex, for something will always get lost in this translation. This "something" that can get lost if we try in too acquisitive a way to reduce higher to lower levels is now often called "information."

The term "information" has technical definitions that need not detain us here. Sometimes it is narrowly defined as the "reduction of uncertainty" and is closely linked to thermodynamic improbability. Here I am using the term more inclusively to cover not only its precise scientific application but also whatever gives order, pattern, meaning and definiteness to things. The point I wish to emphasize is that "information" is a metaphor now used by scientists themselves to indicate their suspicion that much more is going on in nature and evolution than simple exchanges along the mass-energy continuum. In giving definite shape to the evolving world information is logically distinct from mass and energy. It is quietly resident in nature, but in spite of being non-energetic and non-massive it powerfully organizes subordinate elements and routines in the physical world into hierarchically distinct domains.

Michael Polanyi shows us how discontinuous levels of information can exist within what may appear to chemistry, physics or evolutionary science as an unbroken material and historical continuum. Information insinuates itself into nature, he points out, without in any way violating the laws of chemistry and physics. Nevertheless, the discontinuity among various levels of informational pattern in nature quietly fashions a cosmic hierarchy, even though when viewed historically and atomistically nature seems to be a closed continuum with no sharp breaks (Polanyi, 1969).

Marjorie Grene, a disciple of Polanyi, argues that it is possible for us to accept the causal physical continuity in nature's evolution without having to deny that there is ontological discontinuity between distinct "levels" that emerge in the process: ". . . to insist on epistemological and even ontological discontinuity is not to deny historical continuity, for conditions which are continuous can give rise to, or trigger, systems which once in existence are self-sustaining and hence not explicable entirely in terms of the conditions which produced them. The discontinuity of emergence is not a denial of continuity but its product under certain conditions." (Grene, 1961).

There is no logical reason to assume, therefore, that we cannot hold together the notion of nature's atomistic and temporal-historical continuity with that of hierarchical discontinuity. The writing on this page does not violate the chemistry that bonds ink with paper, but the information in my sentences and paragraphs is still logically and ontologically distinct from the chemistry of ink and paper. Similarly, physical or historical continuity does not rule out hierarchical discontinuity in the evolution of life and mind out of "matter."

In its encounter with the informational content in the sphere of living beings science has come up against a kind of causation quite different from what has conventionally been its concern. Modern science has been preoccupied almost exclusively with efficient and material causes, but information is a kind of influence that works non-materially and non-energetically. It powerfully patterns and orders things, but itself remains irreducible to the massive and energetic constituents to which it gives definiteness and identity. Its effectiveness is real, but it slips through the wide meshes of a mechanistic science's net.

The question then arises: where does the information that allows for discontinuity come from? Information, it is important to note, does not originate in the past or evolve out of material stuff in the same way that evolutionists imagine. For in some sense information--in the sense of possible ways of organizing nature--has *always* been there. It has forever occupied an arena that we may refer to as "the realm of the possible."

Moreover, as A. N. Whitehead insists, such possibilities have to reside somewhere (Whitehead, 1978). They cannot arise out of pure nothingness. Even the ultra-materialist Daniel Dennett implies that the informational possibilities that eventually become incarnate in evolution exist eternally in a realm that he calls "design space."(Dennett, 1995). Unlike Dennett, however, Whitehead is a consistent enough thinker to realize that possibilities must be grounded in something *real*. They cannot be identified simply with absolute nothingness, since they clearly contribute to the actuality of things. But where are they? Where do they dwell if they are not always presently available to our scrutiny?

It is partly in response to this question that Whitehead introduces his idea of God as a metaphysical necessity, one without which there could be neither cosmos nor evolution. The informational patterns that order or inform nature cannot originate in bare nothingness, but must come from an inexhaustible realm of possibilities that has always resided in what we might call the mind of God. Firmly grounded in God's transcendent actuality these possibilities are not indiscriminately jumbled but are graded and accorded relative degrees of value by God. Once they have been concretely embodied in nature, in this graded manner, they allow us to attribute a hierarchical status to the evolving universe. The fact that the informational possibilities appear in evolution in the historically gradual, experimental--and always ambiguous--manner that Darwinian science pictures should in no sense, therefore, be taken as logically contrary to the religious intuition of nature's fundamentally hierarchical structure.

It would be of considerable interest, of course, to pursue the question as to why the informational possibilities become actualized in the cosmos only gradually and horizontally rather than immediately and vertically. But it takes only a little reflection to realize that the kind of timeless verticality idealized by the perennialist notion of a sacred cosmic hierarchy is theologically untenable.. An immediate or instantaneous creation of a vertical cosmic hierarchy would imply that God is unwilling to allow the world a share in the liberty and creativity that the best forms of theology have always associated with a truly graced universe. For the world to be genuinely a world unto itself, and not a pure emanation of the divine will, it must be granted the scope to become itself, and this implies *gradually* rather than all at once. The fact that the world cannot conceivably be created all at once--and still be a world unto itself--in turn requires that the universe must always remain relatively incomplete or "unfinished." In other words it would imply a cosmos in evolution. A universe whose creation were instantaneous rather than gradual--or vertical rather than horizontal--would be one no longer open to possibility. It would be a world without a future, and therefore a world without aim or purpose. It is hard to imagine how a finite universe could ever be considered meaningful or purposeful without also being incomplete and evolving. The vastness of time and evolutionary experimentation in this view, therefore, is not a signal of meaninglessness. Rather deep time itself has the meaning of permitting a narrative grandeur in the emergence of hierarchy that a static universe would not allow.

In view of these few ideas of Michael Polanyi and Alfred North Whitehead, it seems rash to suppose, therefore, that the atomizing and historicizing perspectives of contemporary science have in principle destroyed the religious sense of a purposeful universe. Taken together they sufficiently allow that a universe in evolution may still be structured hierarchically--though not vertically or instantaneously--in such a way as to be the repository of momentous meaning and eternal value. Science, of course, is as such unable to detect anything like a cosmic purpose since it leaves out from the start any concern about meaning or value. But this methodological self-restriction is nonetheless capable of being logically reconciled with modes of awareness that *are* sensitive to meaning and value--an awareness that has its chief representation in what we call religion.

REFERENCES

Dennett, D. C., 1995, *Darwin's Dangerous Idea: Evolution and the Meaning of Life,* Simon and Schuster, New York.

Grene, M., 1961, The logic of biology, in: *The Logic of Personal Knowledge,* The Free Press, Glencoe, Illinois.

Haught, J., 2000, *God After Darwin,* Boulder, Colorado.

Lightman, A., and Brawer, R., 1990, *Origins: The Lives and Worlds of Modern Cosmologists,* Harvard University Press, Cambridge.

Polanyi, M., 1969, *Knowing and Being,* University of Chicago Press, Chicago.

Weinberg, S., 1977, *The First Three Minutes,* Basic Books, New York.

Whitehead, A. N., 1978, *Process and Reality,* The Free Press, New York.

Section Four

The Challenge of Sustainability

THE BRIDGE TO HUMANITY'S FUTURE

Willard R. Fey and Ann C. W. Lam

Ecocosm Dynamics, Ltd.
1830 Cameo Court
Tucker, Georgia, U.S.A. 30084-7001

INTRODUCTION

We are all involuntary passengers on a small, open spacecraft. All that stands between us and the frigid vacuum of space is a thirty-mile-thick blanket of air that we must breathe to live and a higher layer of ionized atoms that solar radiation heats to more than a thousand degrees Kelvin. All life on the craft depends on sunlight converted to nourishment by plant photosynthesis. Land plants grow in a thin layer of living soil. Plants absorb carbon dioxide and exhale oxygen, while animals do the reverse. Widespread, diverse plant and animal life forms regenerate the elements necessary for life support and provide flexibility under stressful conditions. This fragile balance of water, atmosphere, soil, and life forms necessary for human survival took more than two billion years to create.

In the last five thousand years, one species alone out of millions has altered the evolution of life on Earth and compromised its life support system. Formerly, all species were created and modified by mutation or long-term natural selection to survive in local environments which created the global atmospheric, soil, and water balances necessary to support all life for eons. Now the local and global balances are being destroyed as the evolutionary process that changed species to fit their local environments is being transformed into the destruction of the global environment to realize the short-term desires of this one species. Humanity is so ignorant of and apathetic toward the details of the automatic natural life support balances that it is threatening its own survival as it allows its acquisitive leaders to become wealthy.

Understanding Complexity. Edited by Ragsdell and Wilby.
Kluwer Academic/Plenum Publishers, New York, 2001

THE ECOCOSM PARADOX

The Ecocosm is the whole Earth system including the planet, the natural nonhuman living systems, and the human presence, all constantly interacting. The Ecocosm Paradox postulates that the environmental crisis arises from the growth of world human consumption, the link between the natural environment and the human presence. Human consumption draws resources from the environment and turns them into waste. When people cut the trees of a forest or pump oil from a well, they do it to provide products for humans to consume. Tragically, the world human system is so dependent on consumption growth that, if growth were to stop to alleviate the environmental crisis, major economic, financial, social, and military disasters would threaten human survival.

History and Warfare

Human warfare developed over time from local, weaponless, nonlethal, territorial combat between individual animals to global nuclear war between allied blocks of nations representing billions of people. The weapons in the early human conflicts were primitive stone spears, knives, and arrows. When horses were domesticated, they were used to pull chariots and carry riders. At sea, wooden, oar-powered galleys used catapults to hurl projectiles. As technology advanced, bronze replaced stone, then iron replaced bronze; guns and cannons replaced catapults, spears, and arrows; and armored vehicles replaced chariots. Galleys were replaced by sail-driven frigates; then by metal, steam-powered battleships, submarines, and aircraft carriers. Recently, atomic weapons delivered by aircraft and missiles brought war to everyone. Now computers, wireless communications, the Internet, and smart, miniaturized, biological, chemical, and nuclear superweapons have created the potential for a species-threatening, pervasive, global killing field with no battle lines. This escalation in the power and extent of human aggression has occurred in an evolutionary twinkling.

The horrors of humanity's military history were demonstrated in many battles that benefited from advances in weapons' technology and changed the course of history: Gaugamela, Actium, Agincourt, Yorktown, Gettysburg, Midway, and Hiroshima. If the environmental crisis created by humanity's growing intervention in the natural life support system cannot be effectively resolved soon; humans will use modern warfare, their ultimate conflict resolution method, to decide the issues arising from water and materials shortages, environmental catastrophes, leadership power struggles, worldwide financial collapse, and failures of the production, distribution, utility, and service functions. Humanity's insensitivity to and acceptance of the massive death and destruction of modern war must end to prevent a violent resolution of the Ecocosm Paradox.

POPULATION

UN data (Figure 1) for world human population are plotted on a logarithmic scale. The exponential shape on the log scale depicts an exponential function whose doubling time is getting shorter. We call it a "hyper-exponential." Human population is now six billion. An additional billion people is added about every 10 years. As the annual growth rate has increased, population doubling time has fallen in the last 500 years from 650 years to its present doubling time of 34 years. The acceleration of population growth is caused by technology-driven health advances that reduce the death rate. Though births per year have fallen also, the world human birth rate is still more than twice the death rate; and major new health advances appear imminent. Strong population growth is not over.

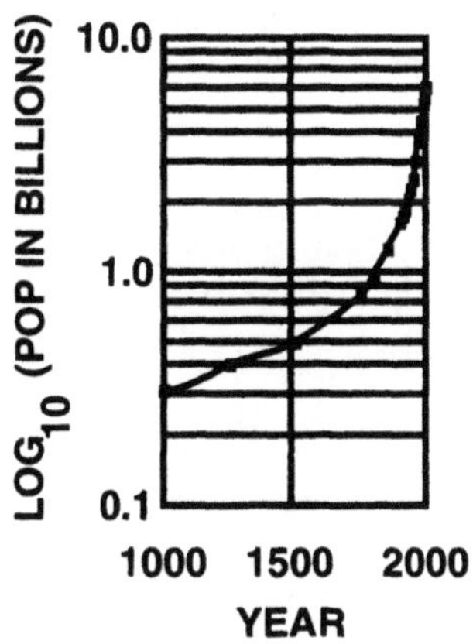

Figure 1. World Human Population Growth (adapted from Parent, 1994)

UN data (Figure 2) show the growth of the world per capita GDP index. This linear time history on the log scale from 1950 to 1980 indicates exponential per capita consumption. World consumption is the algebraic product of average annual per capita consumption and human population. In this graph, per capita consumption doubles in about 30 years. With population doubling every 35 years, their product, world consumption, quadruples in 30 to 35 years. Stopping population growth will not stop consumption growth. If current population growth and per capita consumption growth continue, by the end of the 21st century world human consumption will be 50 times as large as it is today. Since even today's consumption clearly is unsustainable, such growth will devastate the environment.

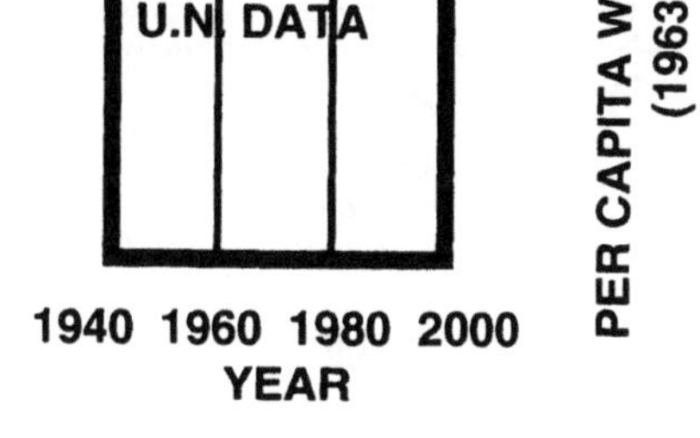

Figure 2. World Per Capita Consumption Growth

Both population and per capita consumption have positive feedback loops that force them to grow (Figure 3). Population growth is driven by a natural reproduction loop and a biotechnology loop that extends life expectancy (Figure 3, right). Per capita consumption growth is caused by technology loops reinforced by world economics and government action loops (Figure 3, left). Environmental collapse, resource exhaustion, and toxic pollution constraints may create negative loops that will finally stop the growth.

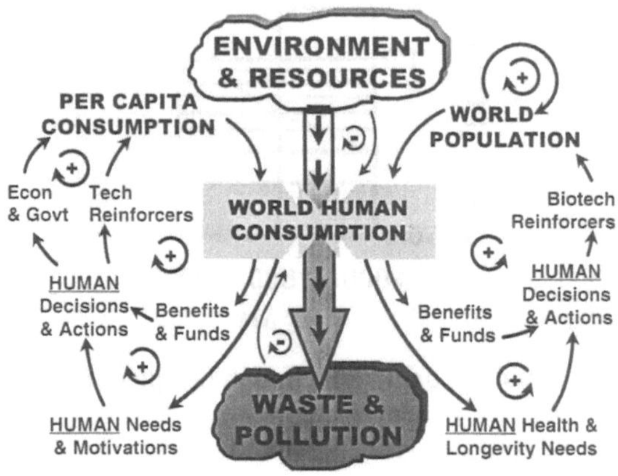

Figure 3. The Ecocosm Paradox Diagram

The causal-loop diagram for human population (Figure 4) shows one positive and five negative loops. Today, human fertility produces new humans faster than deaths and shifts to the non-reproductive state can remove people from the childbearing population that energizes the positive loop. This is the first positive loop that forces consumption growth. In most nations, both birth and death rates are falling. Death rates are falling as public health measures and medical technology extend human lifespan. Funds and benefits from medical consumption motivate biotechnologists to increase research to find more medical products that lower death probabilities even further. This biotechnology loop is the second positive population loop. It makes population's exponential growth hyper.

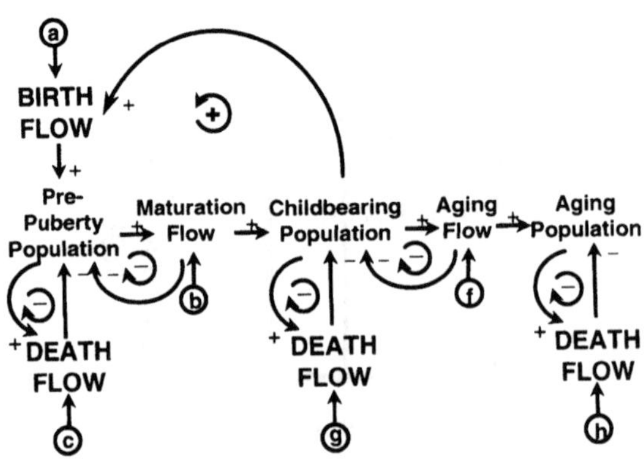

Figure 4. Population Causal Loop Diagram

TECHNOLOGY

Technology (Figure 5) conceives and produces virtually all the products humans consume. Most technological advances result from organized research by highly-trained people working for governments or private organizations. In the upper part of the technology causal-loop diagram, research and design effort leads to new knowledge and products. These motivate higher per capita consumption that increases consumption. Increased consumption of technological products provides many benefits and disbenefits. Both motivate more research. Technology is a positive feedback process in which theories and products developed by scientific research produce results that lead to more research and more trained people to do more research.

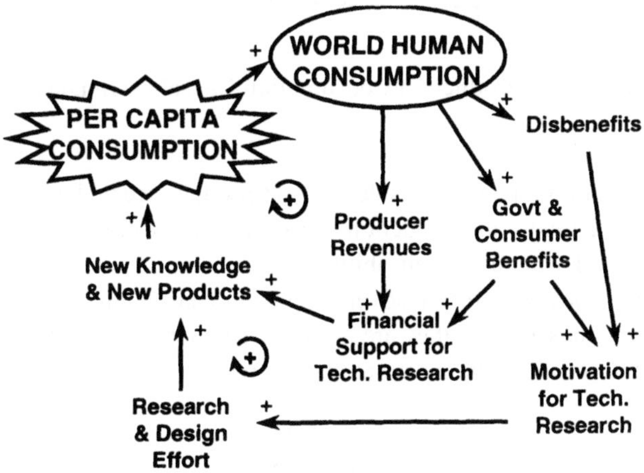

Figure 5. Technology Causal Loop Diagram

ECONOMICS

The interaction of human instincts and technology that drives per capita consumption growth is reinforced by economics and governments. Consumption is the closure point for all major economic loops (Figure 6). It connects the producer and consumer roles of each person and focuses the attitudes and actions of producers and consumers to reinforce consumption. Consumption also relates the financial aspects to the physical aspects of the economy.

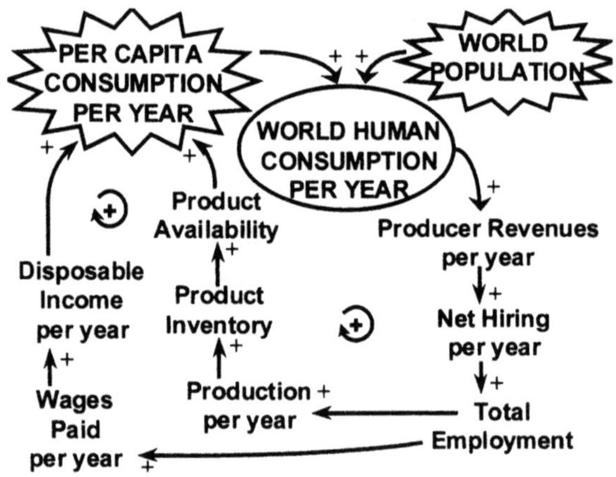

Figure 6. Production/Consumption Causal Loop Diagram

To increase consumption producers design and produce technological products, market them, advertise them, invest in production capacity, hire workers, and provide credit to customers (Figure 7). To increase their ability to consume, consumers work and invest to obtain income. They borrow; perceive advertising that makes them desire more; and use and then discard their purchases when they are used up or better new products appear. Positive feedback loops reinforce all of these actions and, thereby, reinforce consumption.

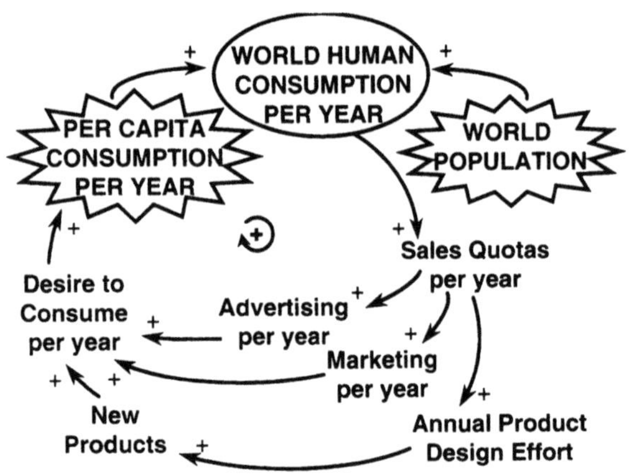

Figure 7. Marketing Causal Loop Diagram

Speculative operations in securities markets are driven by positive investment loops (Figure 8). Consumption growth stimulates price increases for securities that attract speculative funds to finance producer operations that reinforce consumption. If consumption growth were to stop and speculators believed that it would not restart, securities prices would collapse.

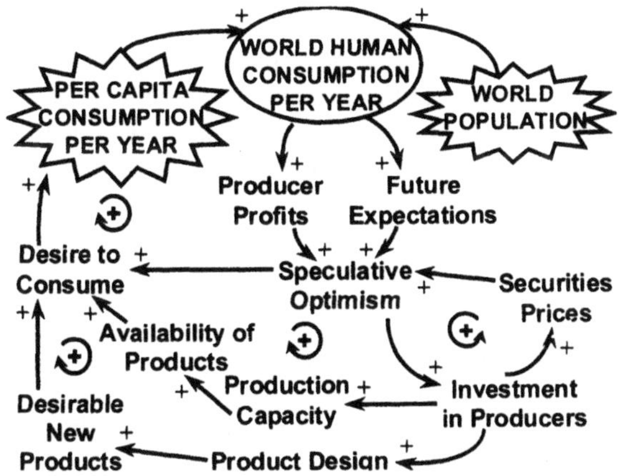

Figure 8. Investment Causal Loop Diagram

Most government revenues come from taxes on income, sales, capital gains, and value added (Figure 9); so, governments encourage consumption growth through their own consumption, the creation of money and credit, depletion allowances and monetary policy. To encourage consumption growth, commercial banks also create money through credit expansion. The operation of positive technology loops reinforced by positive economic, financial and government loops drives the exponential growth of per capita consumption.

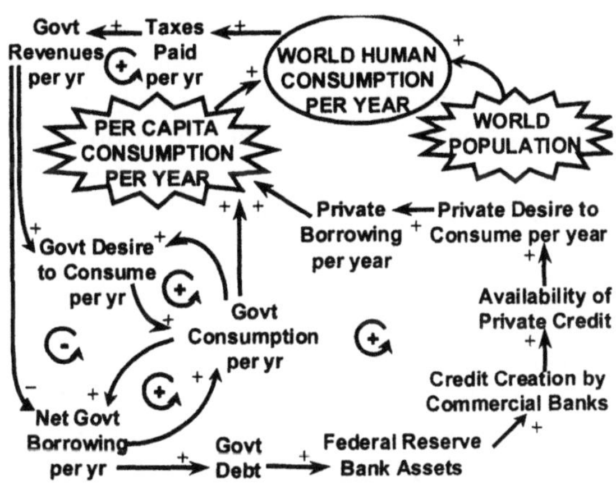

Figure 9. Government Causal Loop Diagram

The Ecocosm Paradox hypothesizes that the growth of world human consumption, which *quadruples* consumption every 35 years, must stop to avoid an environmental catastrophe. A seemingly small 4% per year increase in consumption produces this rapid expansion. Human instincts to reproduce and to consume are enormously amplified by technology, economics, and politics to cause devastating growth. The natural planetary life support system cannot absorb this accelerating destruction and continue to function effectively. The activities of almost all persons, businesses, and governments depend on, create, and demand consumption growth; so it is unlikely that growth can be stopped. But, if, somehow, growth could be stopped to save the environment; societies would become unstable, the world economy and securities markets would collapse, and wars would follow. Thus, humanity's destiny seems to be continued explosive growth until some kind of catastrophe devastatingly ends the growth; and, perhaps, humanity itself.

A NEW APPROACH

No currently proposed solution to the environmental crisis faces the reality that the Earth is finite and human consumption growth must not only stop, but it must be reduced to a long-term sustainable annual amount. This can only be done by changing human attitudes and restructuring the socio-economic system, so it is not dependent on expansion and does not create it. The current piecemeal attempts to improve efficiencies, and to encourage conservation and recycling may postpone the environmental collapse; but they will not stop growth nor produce a sustainable Ecocosm.

A method of analysis, synthesis, and implementation is needed to discover and institute a realizable higher-level human consciousness to support a new, sustainable human socio-economic system structure in interaction with a modified natural life support system. This must achieve a long-term, stable, balanced Ecocosm. It must be in harmony with human instincts, desires, and capabilities; with plant and animal community functions; and with planetary dimensions and resources. It must automatically and quickly limit human consumption, without unacceptable laws or force, to the gradually changing planetary carrying capacity as technology progresses and as resources are exhausted.

The new system must be achievable with "tolerable" damage to human life, freedom, and property. A "painless" solution is impossible. Even today, people are starving and dying in wars. Living systems are extremely complex, internally self-organizing, and respond to depravation and conflict with violence. Humans are not passive and precisely responsive to forces as masses and electrons are. Thus, the success of science in manipulating the physical world is not easily duplicated in the living world, and there is no known algorithm to synthesize a sustainable world human system. Therefore, the solution method proposed here is a science-aided, creative art; not a mathematically derivable, experimentally reproducible science that most scientists would desire.

Designing the adaptive-sustainable Ecocosm and the implementation process necessary to bring it into reality is the most important and the most difficult task ever undertaken in the history of humanity. The method must perform an incredibly difficult analysis to produce a new Ecocosm system paradigm of unprecedented creativity. It can only begin to implement its changes with the cooperation of a substantial fraction of influential human leaders and groups who, in the past have rarely agreed, even on trivial issues. If any of the recommendations are wrong, a catastrophe rather than an improvement may occur. There may be no second chance because changes will have been made that cannot be reversed. Science has produced such toxic materials, powerful forces, and destructive weapons that immense planetary damage can arise from small errors in the planned interventions or from aggression by terrorists, criminals, or warlords. Thus, the method's task is extremely difficult; the time available to do it is unknown, but short; the margin for error is tiny; and the likelihood of a second chance is small. This is not a task that anyone would want to perform unless humanity's survival depends on it. We believe that it does.

The method involves assembling a multifaceted group of people that will create an essential new intellectual discipline as it designs the adaptive-sustainable Ecocosm and then implements the changes in human consciousness and system structure that the design requires. In order to understand and modify the Ecocosm, the designers must be able to conceptualize and analyze the Ecocosm as a whole, living, noisy, feedback control system. While human decisions control the major forces, a diverse nonhuman living community must flourish to provide human life support.

The conceptualization and analysis must include critical aspects of human and living nonhuman behaviors as they interact on local and global levels to preserve the life support system and the living participants. Knowledge from the sciences and other disciplines such as biology, ecology, psychology, theology, economics, sociology, political science, geology, and meteorology will be needed. None of these disciplines was created to study the whole earth system with all of its forms of interaction, especially not in the pathological state of the current environmental and social crisis. Each focuses on a specific, manageable, clearly related set of variables whose properties are isolated from the complexity and incommensurate units of the others. Each is sufficiently complex by itself that it takes years of study for a person to learn the field's principles in depth.

Understanding the Ecocosm will require knowledge from all of these disciplines and others, but seldom in a form the field provides. A new, holistic, statistical-feedback-control discipline is required to combine the essential knowledge of all relevant fields with new knowledge discovered by the group, so the contents of and the relationships between the fields can be applied to the study of the whole Ecocosm as it was, as it is now, and as it will become in the future. The creators of the new field will have to bring available and new information together in one place or reference where it can be obtained from media centers and the Internet. We suggest the field be called Ecocosm Dynamics. Ecocosm is taken from the Greek words "*oikos,*" a house or environment, and "*kosmos*," a complex, orderly, self-inclusive universe. Ecocosm is the whole, orderly, home planetary system of humanity. Dynamics is the analysis and control of patterns of change through time of critical variables.

A group of people is needed to create the Ecocosm Dynamics discipline and to design the adaptive-sustainable Ecocosm. All members will contribute to discipline development and system design. Most will produce their own models and design recommendations. The group will evaluate and consolidate the recommendations and present the best to the world community. In order to accomplish this task, the members must be willing and able to learn about the whole Ecocosm, learn about each others disciplines, share their insights and analysis with the others, and keep open, honest minds. The team must be totally committed to solving the Paradox properly, not to becoming wealthy, powerful, or famous, or to winning prestigious prizes.

It is likely that doing this analysis correctly and honestly will attract animosity from many. The team will be trying to discover a planetary system that will stop consumption growth. Almost everyone is committed to growth and hopes to become rich from it, so no one wants to hear about reducing it. If the team overcomes these obstacles and produces a promising system design, world leaders may not accept it nor allow implementation because the recommendations may seem counter-intuitive to people unfamiliar with the nature of complex, human, feedback control processes. Also, detrimental effects will be expected during the implementation period that few will be willing to accept. If the team's recommendations are approved, any adversity during implementation will appear intolerable to many; and the team will be blamed for it. Once the research work begins, it will soon become clear to the team just how difficult its assignments really are to do properly with reasonable assurance of success in the short time available.

Finding financial support for the group and its tasks will be extraordinarily difficult for similar reasons. Since there is so much at stake for so many growth-dependent people and organizations, both during the final stages of the expansion and in the trade-offs and losses arising from the adversities of the transition period; there will be great pressure on any research group that seeks to influence these outcomes. Those who hold power will not give it up for the good of posterity or for the good of other, less influential, people or life forms. Even if the Ecocosm design is successful, it will only succeed long after the group members are dead. To create the new discipline, to do the Ecocosm analysis, to modify the existing planetary system, and to wait for the transition time pattern to lead to the eventual stable pattern will take at least one hundred years. Only very unusual people will be willing to undertake such a difficult task knowing that their only reward will be antagonism and that they will never know whether they were successful.

As consumption grows faster and faster beyond its sustainable annual amount, humanity will move higher and higher on the precipice above many possible future valleys of disintegration, two of which are shown in the graph of future consumption with and without an implemented solution to the Paradox. The successful development and application of the method; with its research group, new discipline, and design of the sustainable Ecocosm; is symbolized by a bridge over this valley from the unsustainable peak of consumption to a lower sustainable consumption plateau. If an effective design is implemented properly and in time, the Ecocosm will avoid the environmental and social catastrophes that quickly would force the critical variables into the depths of the valley. The design's consumption transition time patterns will also decline; but not as far, nor as fast, with less destruction, and with faster recovery. Therefore, a well-constructed bridge, the method that provides the successful design and implementation, will save many lives and prevent much devastation.

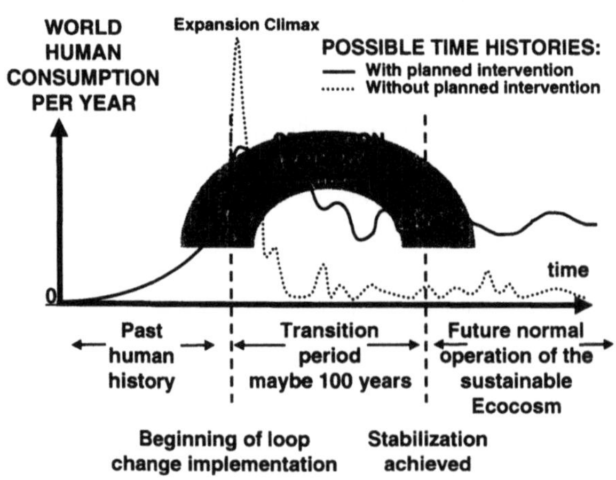

Figure 10. The Bridge to Humanity's Future

We call the effort to design an adaptive-sustainable Ecocosm: Operation Rainbow Bridge. The rainbow, in the traditions of many cultures, symbolizes hope for future well-being after adversity and/or cooperative diversity. It suggests that dishonest and destructive forms of competition will be eliminated; that all of Earth's diverse people will equitably share information, resources and influence; that the adversity of the implementation will be tolerable; that a sustainable Ecocosm will be achieved; and that the result of the reformation process will be to unify all people in their resolve to maintain their stable Ecocosm.

Perhaps the most difficult aspect of the Ecocosm Paradox solution will be the planning and implementation of the transition to the sustainable Ecocosm. In order to stop consumption growth and to obtain long-term stable consumption, certain changes must occur in humanity's consciousness and in the Ecocosm's feedback structure. The design effort must identify these changes. In order to obtain improved performance in the real Ecocosm, these changes must actually be realized in the operating structure.

The Ecocosm's operating feedback structure is created and maintained by the thoughts and actions of humans. Therefore, changes in collective human higher-consciousness must be substantially established before any new loop structure will function properly. The elements of the new higher-consciousness are not all clear; but, at least, humans must overcome their selfishness, vengefulness, and aggression. We must be able instinctively to forgive, to share, and to treat others with honesty, consideration and respect. In a closely interdependent, crowded Ecocosm, trust is the fundamental interpersonal attitude necessary for long-term social survival. Trust cannot persist between people and groups, and the violent resolution of conflicts will not end, unless humanity develops these higher instincts.

When efforts to change consciousness and loop structure begin, transient patterns will arise. These time patterns will be different from old system patterns and different from eventual improved system patterns. Transients will dominate during the transition time period, which begins when structure changes are initiated and ends with the clear beginning of improved patterns as the transients disappear. Changes that are actually implemented are seldom exactly the same as the proposed changes. Unplanned structure changes may be made by system participants who are coping with the system's crises or by those who are trying to evade, circumvent or take advantage of the proposed system changes. Such incorrect, unconscious, and unplanned structure changes may cause serious implementation problems. The result may be chaotic, violent, conflicting, unpredictable, counter-intuitive, and/or uncontrollable time patterns that complicate and lengthen the transition period. At worst, sustainable time patterns may never be achieved.

CONCLUSIONS

On January 16th, 1920 the eighteenth Amendment to the United States Constitution banned the manufacture, sale, and transportation of intoxicating liquors. American reformers wanted to improve people's physical health, family happiness, and social well-being. Instead, consumption of alcohol greatly increased and social stability was upset by the growth of organized crime fueled by bootlegging profits. Thousands died nationwide in the resulting gangland wars and enforcement operations. On December 5th, 1933, the twenty first Amendment repealed Prohibition, but the power of organized crime still remains. Counter-intuitive disasters often result from well-meaning reforms.

Massive human intervention in the natural environment has created Earth's environmental crisis with its Ecocosm Paradox. The intractability of the Paradox and the speed of the environmental disintegration present humanity with a supreme and unprecedented challenge: To create a realizable, collective higher-consciousness that will support a sustainable Ecocosm system structure that will stabilize human consumption, and maintain an adaptable life support system for present living beings and their progeny indefinitely; and to do it before the environment or the socio-economic system collapses. We cannot offer a solution at this time. An effective solution, if one can be found, will be counter-intuitive, even to experienced systems analysts. It is not wise to guess in the face of great complexity with such enormous consequences for humanity at stake.

Others who do make recommendations sometimes have little concept of the true nature and severity of the problems, and little idea of what will be required to solve the problems in the real world. The virtual worlds of television, computers, mathematical abstractions, positive thinking, managed news, psychological advertising, and unprecedented economic prosperity in the midst of escalating environmental destruction have so decoupled science and humanity from reality that the real nature of life on our small planet is almost totally obscured. As Walter Truett Anderson warned in *To Govern Evolution*, "The central political problem of our time is not that people in power do the wrong things, or that some people have more power than others, or that there is a lack of clarity and honesty in political dialogue; all of these are real and serious, but they are only dim reflections of a larger problem, which is that we literally do not know what we are doing."

We have presented here the nature of the Ecocosm, its problems, and the Paradox; with recommendations for finding a solution and implementing it. These are offered as our understanding of the only reasonable chance humanity has to save itself from itself. Virtually all of the prophetic traditions of religions and cultures worldwide, ancient and modern, forecast a catastrophic "end time." Now, a crisis that could threaten humanity's survival is upon us in the form of the intractable dilemma of the Ecocosm Paradox. How will we respond? How will you respond?

REFERENCES

Parent, J., 1994, *A Survey of United States and Total World Production, Proved Resources, and Remaining Recoverable Resources of Fossil Fuels and Uranium,* Institute of Gas Technology, Chicago.

ENERGY, HIERARCHY AND MONEY

Howard T. Odum

Environmental Engineering Sciences
University of Florida
Gainesville, FL, USA.

INTRODUCTION

Energy hierarchy concepts are used to understand the distribution of money. Monetary convergence concept is explained by the universal energy transformation hierarchy. In an energy transformation series, the available energy decreases at each work step, but the transformity, concentrations, turnover times, and spatial convergence increase. These patterns are explained by the natural selection of designs that reinforce their sources. To reinforce effectively with less energy, downstream processes have to be concentrated in space, accumulate, and pulse their feedbacks. Because every real process requires available energy, materials, information, and people organize into pulsing centers. In this essay the distribution of monetary circulation is also linked to the energy hierarchy, providing some quantitative insights on the self organization of an economy. A suggestion is made for a 7th energy law.

We often take for granted the intensive economy in population centers. But these centers may have energetic explanations that follow from the principle of energy hierarchy. Let's consider how the circulation of money is coupled to the hierarchy of energy. Most people would agree that all systems require energy and that economic systems are no exception. But the relationships are perceived as complex and subject to human behavior in markets.

This paper suggests there is a more fundamental, top-down principle: that the circulation of money is distributed according to the hierarchy of energy. In spite of the appearance of unregulated individual actions and randomness, energy transformation rules may control. During self organization, natural selection of human trial and error guides people and their spending into hierarchical patterns because this brings energetic reinforcement.

Understanding Complexity, Edited by Ragsdell and Wilby.
Kluwer Academic/Plenum Publishers, New York, 2001

BACKGROUND

There is a large literature of economic geography, relating the distribution of economic activity, energy, materials, and people (Berry, Conkling, and Ray, 1976). Older literature was summarized by the author's book on ecological and general systems (Odum, 1983, Chapters 17-18, 21). That humans and their cities are distributed hierarchically over the landscape was documented long ago (Cristaller, 1933). The empirical Zipf's Law (Zipf, 1919) found the number of cities and towns inverse to their size. Doxiadis (1977) found increasing concentration of buildings and increasing scale of transportation circulation toward centers.

Anne Marie Jansson and J. Zucchetto (Jansson and Zucchetto, 1978; Zucchetto and Jansson, 1985) made detailed study of the flows of energy, materials and money on the Swedish island of Gotland in the Baltic Sea. While relating energy flows to market values, they postulated that monetary values would increase with embodied energy.

ENERGY HIERARCHY CONCEPTS

In energy transformations, available energy (exergy = potential energy) is used up in processes of work to generate a small output of available energy of a different kind (Figure 1).

So far as we know, all of the energy transformities of the universe can be arranged in a series network suggested by the web in Figure 2a. This is an energy hierarchy because many joules of one level converge to generate a few joules of higher quality.

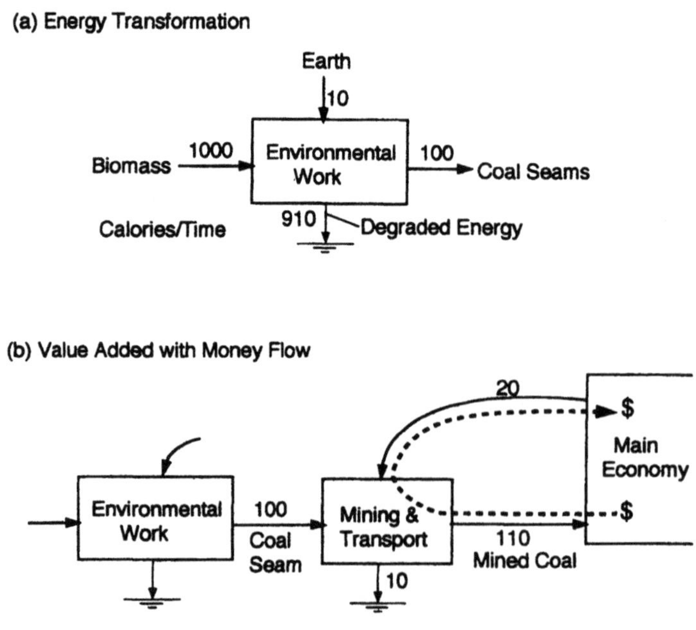

Figure 1. Energy transformed in work. (a) Process without money; (b) a process accompanied by a counter-current of money representing value added.

In most, if not all systems, the higher quality energy feeds back services and materials in control loops to amplify its input process (Figure 2a).

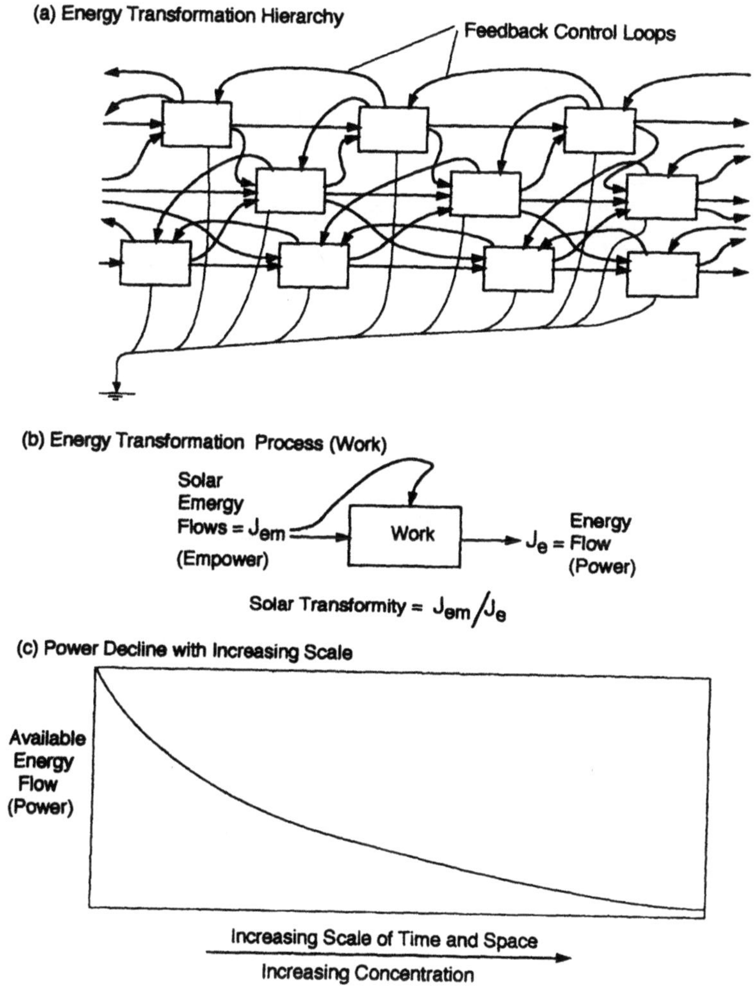

Figure 2. Diagram of the universal energy hierarchy. (a) A network of energy transformations; (b) one energy transformation and the definition of solar transformity; (c) decrease in power through the web of work processes that generates units on a larger scale of time and space with higher transformities.

The available energy flows decreases with each transformation step from left to right in the energy transformation web (Figure 2b), but the scale of the processes increases (Figure 2c). The position in the universal energy hierarchy is indicated by the *transformity*, which is defined as the available energy of one kind required to generate a unit of energy elsewhere in the web. The one kind of energy selected for calculating transformities is defined as the *emergy (spelled with an "m")*. The unit of emergy is the *emjoule*. The *transformity is the quotient of emergy per unit energy*. In Figures 1 and 2 the transformity increases from left to right.

In recent years we used solar energy as the form of energy for calculating emergy. Thus, solar emergy in solar emjoules measures the available energy of one kind which is embodied in any product or service. The flow of useful energy per time is *Power*. Similarly, the flow of emergy per time is defined as empower (Figure 2b).

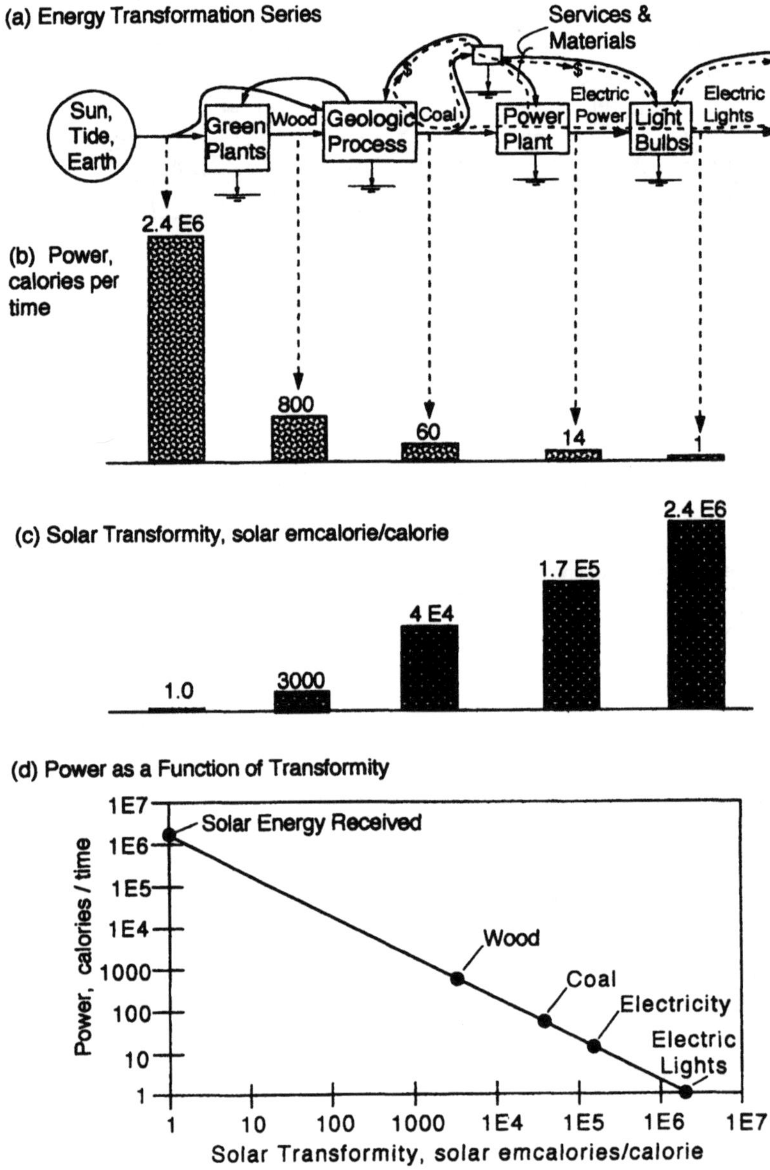

Figure 3. Energy transformation series showing the necessary solar emergy for generating a joule of electric light using electricity from a coal-fired power plant. (a) Systems diagram; (b) available energy used in work in each transformation step; (c) solar transformity of the work products; (d) power used as a function of transformity on logarithmic co-ordinates.

The emergy of something is readily calculated by multiplying the quantity by the emergy per unit from tables summarizing previous studies. For example, multiply available energy by its transformity to obtain its emergy. There is now an extensive literature on emergy valuation, and many transformities have been calculated. The author's book (Odum, 1996) summarizes concepts and earlier work. The first folios of a handbook of emergy valuation containing transformities and emergy values per mass, per person, per bit, etc., have been issued (Odum, Brown, and Brandt-Williams, 2000).

POWER PLANT EXAMPLE

An example of an energy transformation series in Figure 3 connects the solar emergy inputs to the output and use of electricity in a power plant. Because the emergy flow is the same along the series, as the available energy flow (= power) decreases, the transformity increases in inverse relation. At each step the available energy decreases but the solar transformity increases.

PROPERTIES OF THE ENERGY HIERARCHY

The following are properties of the energy hierarchy in space and time as explained in more detail before (Odum, 1987, 1996, 2000). These characteristics result when each product of energy transformation is stored and reinforces its own energy chain with feedback interaction. The natural selection of designs that have feedback reinforcement (maximum power principle) causes the following:

(1) Convergence spatially of energy transformation to form centers (Figure 4) because less energy can have a commensurate feedback effect if it is spatially concentrated.
(2) Energy is accumulated in storages in the centers, delivering its feedback reinforcement in pulses because less energy can have a commensurate feedback effect if it is concentrated in time.
(3) Materials are concentrated in space and time in each transformation step because the available energy which drives material cycles is being concentrated. This means that the quantity of materials concentrated has to decrease in each step because the available energy at each step is less. The emergy per unit mass increases with each step.
(4) The concentration of storages of energy and materials increases with steps along energy transformation series, which is a necessary consequence of using less energy to have a commensurate reinforcement that requires a longer period of accumulation. Examples are reefs in nature and cities in human organization.
(5) The concentration of emergy flow increases since the available energy is being concentrated. Empower per area is called empower density.
(6) Because information requires transformations of available energy to maintain its system of replacement, testing, selecting, and sharing, information is also organized by energy hierarchy with decreasing quantities but high concentrations at higher centers.

Figure 4 summarizes these properties of the energy transformation hierarchy where the emergy flow (empower) is the same through all steps. The properties of power and materials flow decrease inversely to the increasing concentration of energy, materials flow, storages, empower density, and transformity.

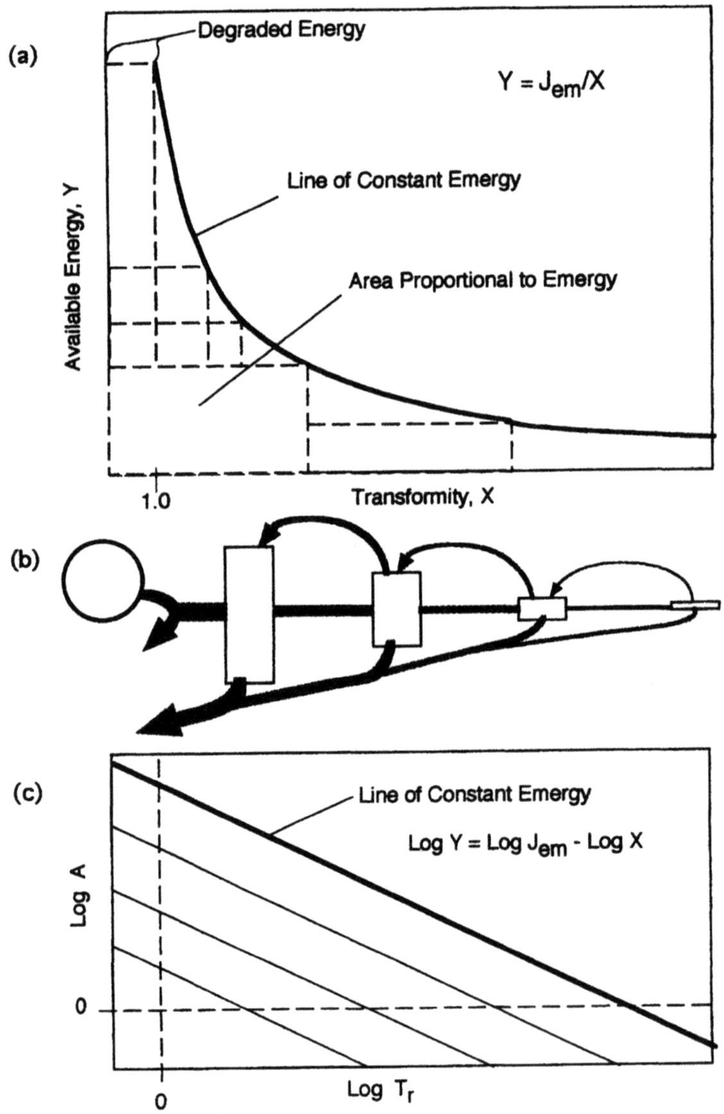

Figure 4. Generalized plot of the energy hierarchy relating the decreases of energy and materials to the increasing concentrations required for commensurate reinforcement feedbacks. Y and X can be any of the following pairs:

Available energy flow and transformity;
Available materials and emergy per mass;
Available materials and empower density;
Useful information available and emergy/unit information.

(a) Linear co-ordinates; (b) energy diagram; (c) logarithmic co-ordinates.

When empower J_{em} is constant, the inverse relationships in Figure 4 are represented with the equation:

$$Y = J_{em}/X$$

where Y is the decreasing quantity, and x is the increasing concentration. Series with many transformations involve so many orders of magnitude that they cannot be represented on regular coordinates (Figure 4a). Data can be readily plotted on logarithmic coordinates. The equation becomes:

$$\log Y = \log J_{em} - \log X$$

a straight line. For example, the power and transformities in the electric power series are plotted in Figure 3d. The energy hierarchy may explain Zipf's law in which city size Y was found to be the inverse square of the city's rank X: $Y = 1/X^2$. The larger the city (hierarchical center), the larger is the territory, and the more energy and steps in the energy transformation series are converging. Because more area is concentrated and accumulated, the concentration is greater. Graphs of cities of different size are straight on a logarithmic plot.

MONEY AND THE EMERGY TRANSFORMATION SERIES

Although we think of money circulation as something people do, each transaction involves an energy transformation. Money flows as a counter current to things it buys. Circulation of money is shown using dashed lines in an energy transformation process in Figure 1b and the power plant example (Figure 3). Figure 5 shows the way additional money flows (value added) with each transformation step toward the hierarchical centers on the right. If money circulation increases with transformation steps (Ts), it increases with the transformity (J_{em}/J_e).

$$J_\$ = k\,(J_{em}/J_e)$$

When the emergy flow (J_{em}) through a series (Figure 5) is constant, then the emergy per unit money ($J_{em}/J_\$$) is proportional to the energy flow (J_e).

$$J_e = k\,(J_{em}/J_\$)$$

The reciprocal of emergy/money is generalized price (price = $J_\$/J_{em}$).
Price = $J_\$/J_{em} = K/J_e$

As the energy decreases in successive transformations, the price of real wealth as measured in emergy units increases.

If area A decreases with transformation steps

$$A = k_2/T_s$$

And money circulation J$ increases with transformations steps T_S

$$J\$ = k_3\ Ts$$

Then the circulation density (J$/A) may increase quadratically with steps.

$$J\$/A = (k\ T_S)/(k_2/T_S) = k_3\ T_S^2$$

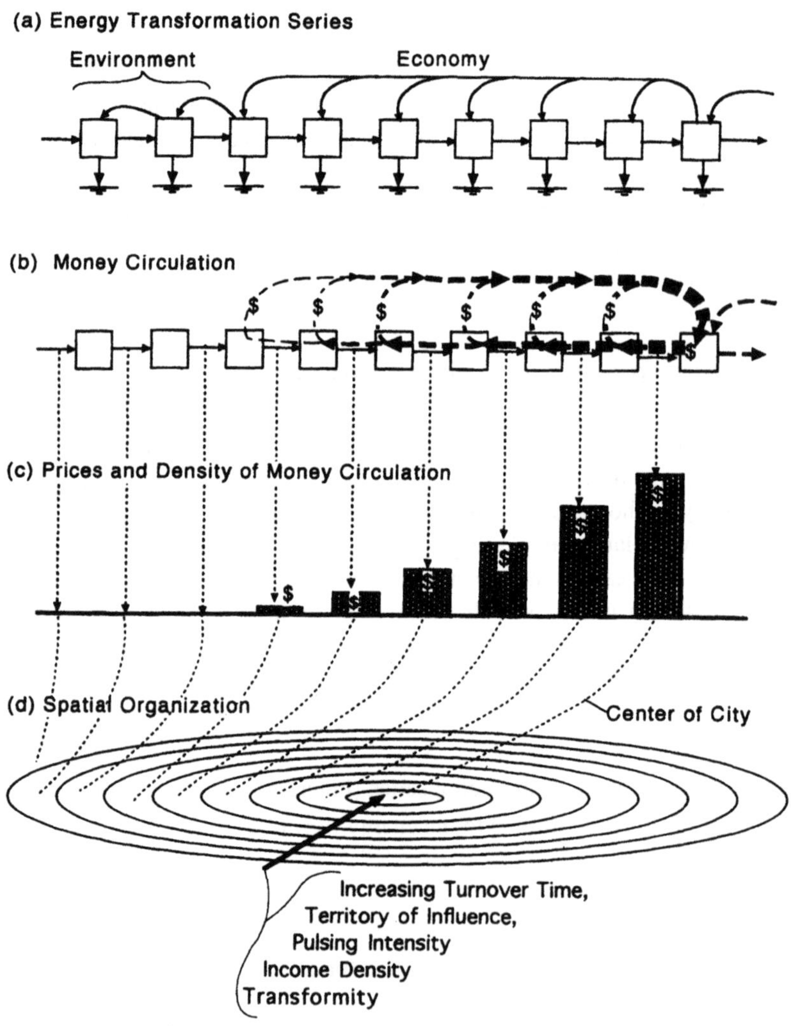

Figure 5. Money circulation in the energy hierarchy with money added at each transformation step. (a) Energy systems diagram; (b) counter current of circulating money (dashed lines); (c) increases in price ($ per unit emergy) with transformation steps toward economic centers; (d) spatial pattern of energy hierarchy with convergence, increased concentrations, and money circulation toward the centers.

Summarizing: money is coupled to energy transformation series (energy hierarchy), and is constrained by the properties of the hierarchy as summarized in Figure 4. Its properties change in passing to higher centers of concentration, the cities (Figure 5). At the low levels on the left are free environmental transformations with no money. At each higher step there is value added, and thus the money circulation increases as does the money/energy ratio. The emergy per unit money decreases, and vice-versa, the money per unit emergy (price) rises. In the centers, the circulation of money is more concentrated but the buying power of money is less. These are familiar patterns that people take for granted. They result from the energy hierarchy principles.

DISCUSSION

The relationships given here link the universal energy hierarchy to the circulation of money. In one sense, energy and money are causally related. Traditional views about the creativity of human choices might regard these coupled relationships as constraints. Others might conclude that human economic behavior, including the free markets, is really not so free of system determinism and natural law.

Whereas the contribution of environmental resources to the economy is not measured by circulating money (Odum, 1986, 1996), money circulating in the markets increases with the emergy and transformity of products once they are above a threshold for economic use and processing within the economy. Thus, this essay supports the Jansson-Zucchetto hypothesis that market values increase with embodied energy (emergy).

In year 2000, with economic growth still in progress, self organization is generating a hierarchy of sub-economies with controversies about the relationship of economic power to other information. The broad base of global energy is supporting high diversity super-circulation. There appears to be a struggle for control between monetary power and the shared information of an emerging world public opinion. The coupling described here helps anticipate the future, as fuel shortages reduce the energy hierarchy, concentrations in centers, and the circulation of monetary buying power. Since emergy is a general measure of inherent real wealth, rising costs of emergy means rising prices.

By recognizing the money-energy connection we can look for other properties expected with energy hierarchy such as increase in period and strength of economic pulses toward the centers. In a reciprocal way, the facts that these concepts seem to explain are also evidence that reasoning based on the maximum power principle is good.

It is customary in science to call a fundamental principle a law after its generality has been widely used and tested. The energy hierarchy concepts are likely to gain this status in time. If we accept the maximum power and empower principle as the 4th energy law, and the energy hierarchy as a 5th, the hierarchy of materials becomes the 6th and the hierarchy of money the 7th. Or perhaps all these will be regarded as corollaries of one energy hierarchy concept.

REFERENCES

Berry, B. J. L., Conkling, E. C. and Ray. D.M. 1976, *The Geography of Economic Systems,* Prentice Hall, Englewood Cliffs, NJ.

Cristaller, W., 1933, 1966, *Central Places in Southern Germany,* Translated by C.W. Baskin, Prentice-Hall, Englewood Cliffs, NJ.

Doxiadis, C. A., 1977, *Ecology and Ekistics* Westview Press, Boulder, CO.

Jansson, A. M., and Zucchetto, J. J., 1978, *Energy, Economics and Ecological Relationships for Gotland, Sweden. A Regional Systems Study.* Ecological Bull. #28, Swedish Natural Resource Council.

Odum, H. T., 1983, 1994, *Ecological and General Systems: An Introduction to Systems Ecology* (Revised edition of: Systems Ecology, 1983, John Wiley) Univ. Press of Colorado, P.O. Box 849, Niwot, CO. 80544

Odum, H. T., 1988, Self organization, transformity, and information. *Science 242* (Nov. 25, 1988):1132-1139.

Odum, H. T., 1996, *Environmental Accounting: EMERGY and Decision Making,* John Wiley, NY.

Odum, H. T., 2000, An energy hierarchy law for biogeochemical cycles, in: *Proceedings of the International Workshop on Emergy and Energy Quality, Sept., 1999,* Center for Environmental Policy, Univ. of Florida, Gainesville. (in press)

Odum, H. T., Odum, E.C., and M.T. Brown, 1997, *Environment and Society in Florida,* Lewis Publishers, Boca Raton, FL.

Whitfield, D., 1994, *Emergy Basis for Urban Land Use Patterns in Jacksonville,* FL, M.S. Thesis, Dept. of Landscape Architecture, Univ. of Florida, Gainesville.

Zipf, G. E., 1919, *Human Behavior and the Principle of Least Effort,* Addison-Wesley, Reading, MA.

Zucchetto, J., and Jansson, A. M., 1985, *Resources and Society A Systems Ecology Study of the Island of Gotland, Sweden,* Ecological Studies #56, Springer-Verlag, NY. 246 pp.

CAPITALISM IN THE NEW MILLENNIUM: THE CHALLENGE OF SUSTAINABILITY

Bill Shireman

Global Futures, Inc. and the Future 500
P.O. Box 160526
Sacramento, CA 95816-0526 USA

ADDRESS

It is a privilege to be here, among so many who have helped to define the discipline of the systems sciences. For many scholars in this field, one of the great challenges of systems thinking has been to find practical applications for the ideas. That hasn't been a problem for us. At Global Futures and the Future 500, our mission is to resolve conflict and forge partnerships between business, environmental, and social objectives. Over the years, we found that in the process of eliminating conflict, we had to think of business and the economy in terms of systems.

So instead of focusing first on the theoretical, and then moving to the practical, we began at the practical and then discovered the theories that were enabling us to solve peoples' problems. Systems thinking worked because of the extraordinary capacity of systems to serve many needs and interests at once. The forest doesn't just maximize for a single output, like board feet. It optimizes across many sources of value. Similarly, by looking at business as a living system, we can optimize our economic, social and environmental impacts overall.

Many leaders in business are discovering this. Five years ago, we brought together 60 such leaders in Aspen, Colorado, to begin to consider how we might operate our businesses sustainably, not just economically, but ecologically and culturally as well. From that meeting emerged a new business network, called the Future 500, composed of companies that sought to identify the qualities that would enable them to be among the 500 most successful companies of the new economy.

The living system emerged gradually as the model and metaphor that has helped enable them to do so. Let me explain how, by relating stories of four business executives this

Understanding Complexity, Edited by Ragsdell and Wilby.
Kluwer Academic/Plenum Publishers, New York, 2001

afternoon who have helped to educate me: Bill Coors of Coors Brewing, Gordon Moore of Intel, Edward Woolard of DuPont, and Tachi Kiuchi of Mitsubishi Electric.

But because I will need to be quick with these stories, let me also invite and encourage you to join in this discussion, at our next annual conference, Industrial Ecology 2000. The date is October 5-8, at the Haas Business School, UC Berkeley. ISSS is our co-sponsor, and let me say how valuable it would be to mix theory and practice together, and see what emerges from the synergy.

I first met Bill Coors in 1985. We were supposed to be arch-enemies. I was head of an environmental lobby seeking to pass a California "bottle bill," and he was the Chairman of one of the largest companies opposing us.

One day I called him, and within a few days found myself in his office in Golden, Colorado. Within minutes we were jointly discussing the issues. That conversation led to the design of the California bottle bill, which is now redeeming more cans and bottles than any other in the nation, with the lowest economic costs and highest environmental benefits.

But I learned something from Bill Coors that was at least as valuable as how to design a better bottle bill. I learned a central thesis of his business philosophy. He told me: All pollution and all waste is lost profit. It's stuff companies pay for but can't sell. One secret to business is to see pollution and waste as a red flag for untapped profits, and seek to reduce, reuse, or recycle it to create value.

Coors applied this philosophy throughout his company. Nothing was spared. Spent grains became fertilizer and feed. Inks that had been used to mark the tops of cans were replaced by ink-free ultraviolet markings that eliminated a toxin and increased production speeds. Wood pallets were reused and then turned into compost. Petroleum solvents were replaced when the company invented a citrus-based alternative, which itself became a new product, Bio-T.

And Bill Coors invented the aluminum can, so that he could buy back his used cans as a source of supply for new ones. He invented not just the aluminum can, but the buyback system of aluminium recycling that still is the economic foundation of much of the consumer-based recycling in the U.S. and Canada.

Finally, he applied the technologies of ceramics used in brewing to develop an array of advanced materials that could be used throughout the economy – in microchips and cell phones and superlight engines – to dramatically reduce the amount of resources needed to get a job done.

The philosophy that led him to do this was the conviction that we had what he calls an "open loop" economy.

In our economy, materials are extracted from the earth, turned into products, and the value is carried through and delivered to consumers. Then we externalize the environmental costs, imposing them on society at large or the earth to deal with. Bill Coors' secret is to "Close the Loop." To draw those materials, or their environmental costs, back to their source. What he's doing is harnessing feedback, the mechanism by which living things adapt.

In nature, when environmental conditions change, organisms can do either of two things: they can adapt or die. The same is true in business. There are two ways to adapt: through efficiency or innovation. Efficiency means making the same product better. Innovation means making a better product.

Edgar Woolard, the former Chairman of DuPont, found that his company could profit by doing both. In 1970, DuPont's chief competitor, Dow, began a famous program to encourage employees to find ways to reduce pollution. In the first few years, hundreds of suggestions were implemented. They paid a return on investment of 173 percent. That means every dollar they spent gave them a $1.73 return in its first year.

Many people assumed that Dow had just found the most obvious sources of waste. Therefore, they thought, the 173 percent return would decline as the company tackled more difficult opportunities. Instead, just the opposite happened. Every idea led to another idea. People found more and more ways to improve. After 20 years, the program was not paying returns of 173 percent anymore. Now it was paying 300 percent. Every dollar brought $3 in savings the first year.

Now let's look at DuPont. They knew that to stay competitive, and also to earn the support of the public and regulators, they needed to match or beat Dow's achievements in eco-efficiency. But they took a different approach. Under Woolard's leadership, they didn't just focus on eco-efficiency. They also focused on eco-effectiveness. Instead of just making the same products more efficiently, they focused on making *better* products - products that performed the same functions as the old ones, but in new ways that were radically more productive.

One of their biggest breakthroughs was in herbicides. At the start of their eco-effectiveness program, DuPont was the world's #7 herbicide manufacturer. They invented new herbicides that were 100 times less toxic than the old ones. They cost less in money, and less in negative health impact. That moved DuPont from the #7 to the #2 position in the market.

When Coors, Dow, or DuPont find a way to make the same product, or create a better one, using fewer resources than before, what is the source of the value they are creating? The source of the value is better design – smarter design – the reshaping of matter into relationship from which emerge greater functions. Gordon Moore, the cofounder of Intel, suggests that we look at the microchip for an example.

The microchip is made mostly of silica – sand – the most abundant compound on earth. What gives it value is not what it is made of, but the pattern that is inscribed on it: its design. Think about this: What is the potential of an economy whose fundamental resource is not physical, not petroleum or steel, but structural?; Whose value is a function of design?

As this resource begins to spread, what companies will be transformed? The simple answer is: almost every company will be transformed by the Eco-Economy. Every company that is now listed as a Fortune 500 or Global 500 company: every retailer, every wholesaler, every manufacturer, and every extractor. Every one of them will either adapt and become part of what we might call the Future 500 (the companies that will lead in the new economy) or they will decline or die and become part of the Fallen 500 (the companies that failed to evolve successfully in the face of change).

Let's look at each of these sectors, one at a time, and see how they are changing already.

Retailers

First, let's look at retailers. Companies like Amazon.com have found ways to sell books, electronics, and other products but eliminate the stores. Are they promoting Green Productivity? Yes. Consumers save 90% of the energy when they ground ship their books instead of driving to the store. They even save half the energy when they have books shipped overnight.

Internet commerce is growing at more than 100% per year. As it continues, what will happen to the world's largest retailers, companies like WalMart, Sears, Kmart, Target, Home

Depot, Kroger, Safeway, and Costco? Will they change and be part of the Future 500? Or will they be replaced by companies with names like Amazon, eBay, and Soma.com?

Wholesalers

Next, consider wholesalers. Home Depot has eliminated intermediate warehouses; 85% of its merchandise goes straight from manufacturers to its stores. Sales associates walk through the store aisles, ordering products that need to be restocked with handheld computers that send the orders over the Internet to 80% of the company's manufacturers.

As the Internet eliminates the need for middlemen, what will happen to the hundreds of thousands of Asian and U.S. wholesalers, agents, warehouses, and distributors? Will they change into new forms, and become part of the Future 500? Or will they decline and disappear?

Manufacturers

Now consider manufacturers. Mass production is slowly being replaced by custom production. Dell Computer is selling computers that meet the specific needs of its customers. Levi's is making jeans designed to perfectly fit specific buyers. MP3.com is allowing millions of music buyers to make their own CD's, with just the songs they want, by downloading them directly off the Internet. What will happen to the world's mass manufacturers? Will they change to serve their customers in new ways? Or will they decline, and become part of the Fallen 500?

Extractors

Finally, consider the extraction companies: companies that take resources from the earth, like oil companies, mining companies, timber companies. Today, Exxon, British Petroleum, and Mobil are the world's 8th, 19th, and 40th largest companies. Yet demand for oil is flattening. In the U.S., the economy grew 9 percent in 1997 and 1998, yet energy efficiency grew 4 percent each year. Overall, demand for energy stayed the same. That means we got 9 percent more economic output, using zero additional energy.

Now, with oil prices higher than ever, many think the oil companies will grow in power. But in my opinion, this will be short term. Today it is cheaper to save energy than to pay high prices. That means, just like in the 1970s, energy efficiency will grow in the years ahead. The demand curve for oil will decline, prices will fall again, and the oil companies will be forced to adapt or decline. Will they change and become part of the Future 500, or stagnate and join the Past 500?

That depends on their capacity to excel in the midst of change – to take on the three characteristics that Lane Tracy says are typical of living things – to sustain, develop, and propagate themselves. So the living system becomes the new model for the dynamic, adaptive, resilient business of the future?

What model of life can we use to illustrate those qualities? What model can instruct us about how to manage our businesses, our economies, and lives? I learned that model from Tachi Kiuchi, the former chairman and CEO of Mitsubishi Electric America. Kiuchi looks to the forest – especially the rainforest – for his lessons about how to operate businesses sustainably.

Yesterday, in the old economy, Kiuchi says, we managed business as a machine, a top-down hierarchy to quickly extract resources from the earth and turn them into billions of products, all the same. Today, in the Internet economy, no business can excel if it is managed like a machine. It must learn to operate like the systems of nature - to be adaptive, creative, resilient, and sustainable.

There is a secret to managing a business like a living system. It is a secret that every gardener knows. The secret is this: you cannot manage the business the same way at all times. Instead, you must manage it in four different ways - a different way for every season.

Every living thing -- every garden, every business, every person in this room -- evolves through four seasons, four phases of life, over and over again. Seasons of spring, summer, fall, and winter: creation, growth, development, and renewal. The essence of sustainability, in nature and in business, is learning to prosper not just in one season, but in all seasons - to complete the cycle of creation, growth, development, and renewal. Here is how to excel in each season.

To excel in the spring, the season of creation, we must plant many seeds, many ideas, and many innovations.

To excel in the summer, the season of growth, we must select – we must choose and nurture a very few seeds - those seeds best matched to their environment, those seeds most capable of extraordinary growth.

To excel in the fall, the season of development and challenge, we must adapt – we must respond to limits by learning to be thrifty - to improve our efficiency and our quality, our degree of perfection.

And to excel in the winter, the season of destruction and renewal, we must preserve our seeds, our templates, our DNA, some the same and some different, from which to begin a new phase of creation and growth.

The companies that excel today are those who structure themselves as adaptive, resilient, creative, and sustainable – living companies with the capacity to learn and change. These companies are increasing their profits in new ways: turning wastes into new savings and products, leasing and remanufacturing billions of dollars of equipment that used to be thrown away, driving pollution and waste toward zero, and systematically eliminating products and even whole industries while continuing to provide the services that are what we actually want.

These are companies that know how to excel in the season of creation. Like 3M, one of the oldest major corporations in the U.S., a third of whose profits come from products introduced in the last five years. Like Xerox, which invented the first personal computer, the first intranet, the first email, the Macintosh-style interface – yet failed to do what it had to do to profit from those innovations – to select.

Xerox failed to do what other companies know is essential to succeeding in the season of growth. To select those seeds best suited to growth.

Like Ford Motor Company, which early this century mastered the machine management principles of the great Frederick Taylor, the father of scientific management, then made quality its focus in the latter part of the century, and is prepared even to move beyond the internal combustion engine to a new product line more reflective of its purpose in the new conditions of the century ahead.

Like Interface Carpets, which grew to be the world's largest maker of commercial carpet tiles like the ones in this room, and is now investing its profits to excel in what it calls the next industrial revolution, as a company that provides the services of floor covering, without needing to sell the product itself.

Like Nike, which achieved its goal to make the world's best footwear, and grew to become the world's largest as well, then survived challenges to Reebok and others to enjoy whole new sales surges, and now has set its sites on becoming a sustainable company.

Like MacMillan Bloedel, recently purchased by Weyerhaeuser, which grew to become Canada's largest forest products company by consuming the trees of old growth forests, and now is setting aside vast regions of remaining ancient forests for preservation, and learning new methods of forestry that retain the ecological integrity of the forests.

Companies that know how to adapt, are applying the principle for success in the season of development. They manage themselves as learning organizations, seeking continual improvement, and measure the efficiency of their operations in terms of energy, resources, capital, and labor.

Like Target Stores, which uses Activity Based Costing to measure the waste intensity of every one of its stores, and through this catalyst triggers hundreds of efficiencies that together dramatically streamline the operations of its stores and its suppliers, saving money and resources.

Like Texaco and Shell, who in Bakersfield, California have turned an oil refiner into a complex industrial ecosystem, whose wastes from one process become money-saving resources for others.

Like Coors, which sees all waste as lost profit, and has recycled its wastes into everything from fertilizer and feed to aluminum cans, and recycled its ideas to create through its spin-offs more than 10,000 products whose value comes from their advanced design.

And there are companies that have excelled through the most challenging of all business stages: creative destruction and renewal.

Like Royal Dutch Shell, whose visionary leaders like Arie DeGues first invented the term "living company," and applied it to navigate Shell through periods when petroleum was challenged more fundamentally than ever before.

Like Mitsubishi Electric, which provided the electrical equipment that helped Japan's economy grow early in the century, led in the electronics revolution of the middle century, and is now pioneering the ideas of industrial ecology that can foster growth in the living economy.

Now through the Internet, a new generation of companies are emerging that can provide for virtually any material need beyond food and housing using a tiny fraction of the resources once required. These businesses are systematically eliminating the need for the products and processes that used to be required to deliver the services that we truly desire.

The key to business success in the new economy – the key to being among the Future 500 – is to systematically eliminate the need for the resource-intensive, environmentally destructive products and processes of the industrial era. Look to the industries that involve the most damage to the earth, and aim to take their market share away, by creating more efficient products that reduce environmental costs, and more effective products that radically reduce costs. That is the objective of the companies that will be joining together at our conference October 5-8, 2000 at UC Berkeley. I invite each of you to join in that discussion.

Today, as we approach the end of the old industrial machine age, it is time to move beyond machine styles of business, and the materialism they foster, and adopt new systems of management and measurement, and a new set of values, that see businesses and the economy as living systems, sustained by profit, performance, and an underlying sense of purpose, long into the future.

Section Five

The Art and Science of Forecasting in the Age of Global Warming

TECHNOLOGICAL FORECASTING AND ASSESSMENT TOOLS: THE CHANGING SCENE

Harold A. Linstone

Past President, ISSS
Portland State University
P.O.Box 751
Portland, OR 97207-0751

LOOKING BACK

World War II constituted a major discontinuity in technological forecasting (TF). Before the War, U.S. research and development budgets were in the millions; after the War they soared into the billions. Technology, exemplified by nuclear weaponry, guided missiles, radar, computers, rocket engines, and supersonic jet aircraft, assumed critical importance in the Cold War. The unprecedented complexity of these systems required intensive development and use of systems analysis techniques.

The perceived Soviet threat and long lead times for complex systems made it vital for the military to forecast technological possibilities further into the future than ever before. This need led the American military and its aerospace contractors in the 1950s and 1960s to ask their systems analysts to focus on quantitative technological forecasting tools. They included exploratory techniques (proceeding from past to future capability) such as trend extrapolation, S-shaped growth models, and precursor analysis. There were also normative approaches focusing initially on future needs analysis. An effective forecast must, of course, combine both capability and need. Scenarios and Delphi proved particularly popular qualitative tools.

During the same time, in the Soviet Union an entirely different approach relevant for technological forecasting was developed. It was based on an analysis of hundreds of thousands of worldwide patents by G. Altshuller to deduce patterns of technological innovation and laws of technological system evolution. The work, known by the acronym TRIZ, was published in Russian and for decades was little known in the West.

In the mid-1960s, as a result of books such as Rachel Carson's *Silent Spring* (1962) and perceived concerns about the potential adverse impacts of technological systems like the proposed supersonic transport with its shock wave, a need for "technology assessment"

(TA) was recognized. Now systems analysts were forced to go beyond purely technological aspects to deal with far more difficult social and human factors and their tool kit proved quite inadequate. For example, there was frequent "assumption drag", that is, the explicit or hidden societal assumptions in a forecasting model were reasonable at the time the forecast was made, but proved in hindsight to be unsuitable for the future time period, thus leading to poor forecasts.

Beginning in the 1970s the recognition of the limitations of systems analysis for planning, as well as the easing of the Cold War, signaled a shift away from the development of technological forecasting and assessment in the West. The change was exacerbated by the fact that the non-military private sector and the general public tended to focus on the near-term future. Their planning horizon was much shorter than that of the military and aerospace community with their long lead time systems. This characteristic of discounting future problems and opportunities relative to the present is known as "the discounting dilemma" (Linstone, 1999).

LOOKING AHEAD

The difficulties encountered challenged technologists to consider new approaches to forecasting and assessment and seeds began to sprout which augur a new generation of tools. We shall briefly sample them.

Multiple Perspectives

This concept is a direct consequence of the perceived limitations of systems analysis. Based on the work of Graham Allison (*Essence of Decision*) and West Churchman (*The Design of Inquiring Systems*), it augments the standard systems analysis perspective, now labeled the Technical Perspective (T), with Organizational (O) and Personal (P) Perspectives. Table 1 outlines some of their distinguishing characteristics. For example, in the case of the *Exxon Valdez* Alaska oil spill, organizational perspectives not only included Exxon, but the US Coast Guard, the State of Alaska Department of Environmental Conservation (ADEC), and the Cordova District Fishermen United. Personal perspectives included key individuals, such as Exxon's chairman, the ship's captain, and one of the experts at ADEC. Each of these three types of perspective uses distinct paradigms and each provides insights not attainable with the others. Their synthesis or integration bridges the gap between the systems analyst or modeler and the real world. It is thus particularly appropriate for technology assessment and its first applications, funded by the National Science Foundation in the late 1970s and 1980s, were precisely in the TA area and proved the value of the approach. It has also been applied to forecasting, for example, exposing assumptions that bias forecasts and clarifying scenario types. (Linstone, 1999)

Complexity Science

In the 1980s an interdisciplinary team at the Santa Fe Institute developed a new computer simulation approach to the analysis of complex sociotechnical systems. These nonlinear adaptive systems are found to exhibit various states, (1) stable, that is, converging to an equilibrium, (2) oscillating stably, (3) chaotic with predictable boundaries, or (4) diverging unstably. In the chaotic state the system seems to show paradoxical behavior: it is deterministic because it is fixed by equations and yet it may give an appearance of randomness. It may be orderly and suddenly become chaotic or vice versa. The system is exceedingly sensitive to initial conditions, making the use of historical data as a basis for forecast dubious at best. Adaptiveness means that each system element cannot see the whole picture but has its own internal models, that is, its own perspective. It must base its decisions on local information, but has the ability, using feedback, to create or revise the models or rules governing its actions. The total system emerges from the self-organization of its parts and thus is not optimizable from the top down. The concept underscores the point made by the multiple perspective approach: "rational" organizational or individual behavior of system components is by no means equivalent to rational total system behavior or optimization. Most significant of all is the recognition of inherent limits of predictability - the chaos phase. In simulations of primitive exchange-type economy models with thousands of individual trading elements or agents, it was found that some foresight on the part of the agents is better than none, but large amounts of foresight are less "fit" than modest amounts (Epstein and Axtell, 1996). It is also useful for the long-range planner to keep another characteristic of complex adaptive systems in mind: they continually evolve and do not have optimal end states.

Table 1. Some characteristics of the three multiple perspective types

	Technical (T)	Organizational (O)	Personal (P)
World view	Science or technology	Group or institution	Individual or self
Objective	Problem solving	Action, process, stability	Power, influence, prestige
System focus	Artificial construct	Social, institutional	Genetic, psychological
Ethical basis	Logic, scientific rationality	Justice, fairness	Morality
Mode of inquiry	Observation, analysis, optimization, modeling, cause and effect	Consensual, bargaining, satisficing, agenda, problem delegation	Intuition, learning, challenge and response, creativity by the few
Planning horizon	Far	Moderate discounting	High discounting for most
Uncertainty view	Uncertainties noted	Uncertainties used	Uncertainties disliked
Communications	Technical report, briefing	Insider language	Personality, charisma vital
Criteria for acceptable risk	Logical soundness, openness to evaluation	Institutional compatibility political acceptability	Conduciveness to learning, time-space distance of event
Scenario type	Probable	Preferable	Possible
• criterion	analysis (reproducible)	value (explicative)	image (plausible)
• orientation	exploratory (extrapolative)	normative	visionary
• mode	structural	participative	perceptual
• creator	think-tank teams	stakeholders	individuals

It is now recognized that the beginning and end of the familiar S-shaped growth curves are domains of instability and chaos. This accounts for the fact that the "take-off" point of an innovation and the shift from one S curve to the next present barriers to the forecaster, whereas the smooth stable growth portion between them is a realm of reasonable predictability (Modis and Debecker, 1992). The bifurcation concept in chaos theory may also prove relevant in the analysis of innovation patterns.

Potentially, this approach can prove insightful for TF/TA. Here are some hints:

- The role of randomness in innovation is vital; it creates fluctuations that act as natural seeds from which new patterns and structures grow. It is a point well known by managers: shake up a stable organization if you want to foster creativity. Mapping the domains of stability and chaos may permit facilitating chaos in a stagnant system.
- At times it may be desirable to delay or forestall a phase change. For example, it may be dangerous to speed up information flow when there is the potential of inducing chaos that management cannot handle. Inappropriate timing of the onset of chaos can be averted by cutting feedback loops in the system. On the other hand, improving feedback can enhance the agents' local information, emergent self-organization and the bottom-up decision making process.
- Work with such models (Epstein and Axtell, 1996) can develop insights on critical questions posed by the impact of technology on society. What is the desirable balance between organizational centralization and decentralization in the corporation? How should the governance of networked systems evolve in the age of globalization?
- Considering the insights provided by the simulation approach to biological system evolution, the intriguing question arises: can artificial technological worlds be created in the same manner, obeying the same laws? This might make it possible to view the technological web as driving its own transformation by continual creation of new niches. Goods and services invented provide new opportunities to create still more goods and services, expanding the technological frontier. This process could be varied to determine sequences with the highest value added (Kauffman, 1995).

Coevolution of Technology and Organizational Networks

Rycroft and Kash (1999) have examined complex technological products and processes and have found in the six case studies considered that they all involve coevolution of technology and continually adaptive private/public organizational networks. As the technology evolves, so do the organizational arrangements. Thus the cardio-imaging concept initially involved phased array technology at Hewlett-Packard Laboratories and Stanford University, then color flow imaging at Duke University Medical Center and the National Oceanic and Atmospheric Administration (NOAA), then integrated backscatter technology at the University of Massachusetts and Duke. In no case did one organization develop the innovation on its own. The implication is that TF, like TA, will require multiple (T and O) perspectives. It can be meaningful for complex technological systems only if it recognizes, and becomes part of, the coevolutionary setting that will dominate in the new century.

Bibliometric Analysis and Data Mining

It may well be that technological intelligence rather than accurate trend extrapolation, modeling, or Delphi will prove most useful in enhancing foresight. Accessing the wealth of electronic databases for detection of emerging technologies and processing both mainstream and outlier pieces of information may yield the keys to new combinations of technology. We recall that the sea-launched ballistic missile Polaris required at least four distinct technological innovations: nuclear propulsion systems for the submarine, solid propellant rocket fuel, a precise inertial navigation system, and a heat-resistant reentry nose shield. The breakthrough required technology transfer and technology fusion.

Porter (1999) suggests the microchip to advance genetic technologies (the genechip), chemical applications (microfluidics), and micro-electromechanical devices (MEMS). A hypothetical example: (1) using microfluidics to present environmental samples to (2) DNA-based sensors that identify particular toxins leading to (3) actuation of MEMS-based devices that counter the threats posed by the toxins. The technology rarely mentioned in conjunction with the focal technology may convey the critical knowledge of a novel linkage of technologies. Thus the clever use of massive electronic databases can mine information and tease out relationships that elude conventional technological forecast and foresight techniques. Clues to imminent industrial disasters, for example, hidden failure modes as well as unintended consequences and impacts may also be uncovered.

TRIZ

The further development of TRIZ is under way with emphasis on directed technology evolution. This normative approach is a formalized procedure that allows for the proactive identification of a strategic objective and development of tactical plans for achieving it. Clarke (2000) illustrates the approach with the case of endoscopic surgical instrumentation. First, the evolution of the existing systems for wound closure, sutures and mechanical devices such as staplers, is traced. TRIZ has found that, when moving towards "ideality", one pattern of evolution proceeds by transitioning to a microlevel to achieve more universality. In this case it suggests that tissue interaction would be effected by a formless liquid or gel that could take any shape. Another pattern of evolution, increasing complexity followed by simplification, is applied to the sewing task. The benefits of sewing and stapling are combined in a microlevel lattice holding the tissue together. A next-generation instrument combining these two patterns of evolution is the result: a reservoir containing liquid polymer with a nozzle connected to a pressure source and the polymer extruded in a narrow knife-like stream that pierces the tissue and then solidifies in a continuous V-type lattice form. The new system eliminates needles and sutures. In this systematic exploration of technologies, many other possible paths of technological evolution can be examined.

It should be noted that TRIZ has aspects of the morphological approach suggested by Zwicky in the 1940s but has more normative aspects. Another morphological approach is being developed by drug companies: using computers to try vast possibilities of genetic variants to search out effective new drugs.

Scenario Management

As an example of significant improvement of existing tools, consider scenario development. Gausemeier and associates (1998) have been developing a computerized scenario management process that is a considerable advance over previous procedures and particularly adapted to entrepreneurial business decisions. It allows the inclusion of company-specific organizational perspectives in developing strategies. Starting with an assessment of the decision field (market share, distribution, profit, functionality, service,...), scenario fields are developed. These encompass 60 to 150 influence factors, drawn from areas such as the industry (competitors, customers, suppliers,...) and the global environment. External and internal scenarios can be created and an influence matrix helps to select the key factors. Their possible developments are projected into the future and these are then grouped into a few contradiction-free scenarios using cluster analysis. For example, for an ATM manufacturer the key factors would include buying habits, home shopping, cashless money transfer, and communication costs. The consequences of these resulting scenarios for the decision field are determined. Alternative strategies (planning oriented, opportunity seeking, and proactive) can then be evaluated in terms of a single scenario focus or robustness based on multiple scenarios.

The Discounting Dilemma (Linstone, 1999)

Perhaps the most profound issue for forecasting is the human predilection to apply a discount factor to the future (as well as the past). The further an event lies in the future, the less we are interested in it. Thus global warming and the disappearance of biological species in the 21st century are seen as remote events and do not spark strong public concern. In other words, we view the future as if through the wrong end of a telescope: the farther away the object, the smaller it appears. It is a concept familiar to economists, who apply a discount rate to future income and expenditures, but it is a universal human characteristic, quite understandable when uncertainty inevitably increases as we peer into the future. The discounting of distant events is not only found in the time dimension but also in space: we are more concerned with happenings in our neighborhood than in physically distant places. The situation is exacerbated in the information technology era, as time and space dimensions are shrunk in the global village. What can be done?

Several paths can be pursued. Media such as radio, television, and film offer highly effective means to bring the distant future nearer. Recall the widespread response to the 1936 radio broadcast of H. G. Wells' *War of the Worlds* by Orson Welles' Mercury Theater. Computer games such as SimCity and SimEarth constitute tools that can increase the understanding of complex adaptive systems. We also observe that telecommunications can bring distant physical space to near space particularly well: recall how we could see the manned landing on the moon instantly in our living rooms. If it is possible to transform a distant-time/near-space event into a near-time/distant-space event, we can then visualize its effect much more readily. For example, future prospects of social legislation (care for the aging, paternity leave) for the U.S. can be examined by using the media to look at the implementation and effects of current social legislation in Scandinavia, which tends to lead the U.S. by about 15 years. Similarly, future environmental destruction in our neighborhood can be made vivid by examining precursor effects seen in other global areas today.

Crisis Management

Finally, the inherent limitations of forecasting noted make it imperative to develop crisis management capability. Moreover, advancing science and technology are revealing new realms of ignorance and compounding the difficulties of technology assessment. Boorstin (1994) has labeled our times the Age of Negative Discovery. More surprises and inconceivable events are inevitable and training to handle the unexpected must be enhanced.

The military train their officers in many scenarios (war games), not with the expectation that one of these will materialize, but in the hope that they will have internalized the ability to react effectively in an unfamiliar situation. As Karl Popper argues, logically present knowledge cannot know the contents of radically new knowledge. Harvey Brooks (1974) talks of "the impossibility of proving impossibilities", while Arthur Clarke

(1962) observes: "When a distinguished but elderly scientist states that something is possible, he is almost certainly right. When he states that something is impossible, he is very probably wrong".

For the decision maker a useful analogy is the task of steering a boat in a turbulent stream. There is only partial control and the view ahead is often obscured. The keys are constant alertness and quick reaction, steering clear of looming rapids, rocks, and shore, in other words, catastrophe avoidance (Dror, 1999).

The importance of TF/TA in this technology-driven economy is clear. This brief survey suggests that its tools will benefit from the fresh thinking under way and new procedures will evolve to facilitate the tasks.

AN INTERPRETATION

A significant comparison can be made between the changing scene of TF/TA and the Kondratiev long wave pattern. Each 50-60 year economic cycle (prosperity-recession-depression-recovery) exhibits a 25-30 year upswing and a 25-30 year downswing. The upswing is a time of knowledge consolidation and steady exploitation of available technology; the downswing a time of creative destruction (Schumpeter's term) and knowledge innovation. Thus it is not surprising that we find a cluster of major technological innovations -- computers, nuclear power, television, radar, jet engines -- evolving in the Great Depression. Following World War II we experienced the 25-year upswing of the 4th Kondratiev wave, which saw the flowering of systems analysis and quantitative technological forecasting tools and the rise of think tanks such as The RAND Corporation. Near the peak of the cycle (around 1970), the forecasting tools were formalized in a number of textbooks (Jantsch, Ayres, Martino, etc.) and journals such as *Technological Forecasting and Social Change* and *Futures* were started. However, in parallel with the start of the downswing, disillusionment with systems analysis, forecasting, and planning soon set in. Typical magazine headlines were "Corporate Strategists Under Fire" (*Fortune*, December 27, 1982); "The Forecasters Flunk" (*Time*, August 27, 1984); "Goodbye, Corporate Staff" (*Fortune*, December 21, 1987).

But this era of creative destruction also fostered knowledge innovation and risk taking, leading not only to another cluster of new technologies -- the internet, genomics, nanotechnology -- but also nourishing the methodological seeds described in the preceding section. Whether the trough has been reached with the fall of the Soviet Union and the Berlin wall, as well as the Asian economic crisis, remains to be seen. If the pattern persists, we anticipate a surge of new tool development based on these methodological seeds during the 25-30 year knowledge consolidation upswing of the 5th Kondratiev wave in the 21st century.

REFERENCES

Boorstin, D. J., 1994, *Cleopatra's Nose: Essays on the Unexpected*, Random House, New York.

Brooks, H., 1976, Herbert Spencer Series on Technology and Society lecture, University of Pennsylvania, April 14.

Carson, R., 1962, *Silent Spring*, Houghton Mifflin, Boston.

Clarke, A., 1962, *Profiles of the Future: An Inquiry into the Limits of the Possible*, V. Gollancz, London.

Clarke, Sr., D. W., 2000, Strategically evolving the future: directed evolution and technological systems development, *Technological Forecasting and Social Change*, 64:133.

Dror, Y., 1999, Beyond uncertainty: facing the inconceivable, *Technological Forecasting and Social Change*, 62:151.

Epstein J. M., and Axtell, R., 1996, *Growing Artificial Societies*, The Brookings Institution, Washington, D.C., and The MIT Press, Cambridge, MA.

Gausemeier J. et al., (1998), Scenario management: an approach to develop future potentials, *Technological Forecasting and Social Change*, 59:111.

Kauffman, S. (1995), *At Home in the Universe*, Oxford University Press, New York.

Linstone, H. A. (1999), *Decision Making for Technology Executives: Using Multiple Perspectives to Improve Performance*, Artech House Inc., Norwood, MA.

Modis T. and Debecker, A. (1992), Chaoslike states can be expected before and after logistic growth, *Technological Forecasting and Social Change*, 41:111.

Porter, A. L. (1999), Depth perception, *Technological Forecasting and Social Change*, 62:143.

Rycroft R. W. and Kash, D. E. (1999), *The Complexity Challenge: Technological Innovation for the 21st Century*, Pinter, London, UK.

FORECASTING THE INNOVATION OF COMPLEX TECHNOLOGIES

Don E. Kash and Robert Rycroft

George Mason University
School of Public Policy
4400 University Drive, MSN 3C6
Fairfax, VA 22030

INTRODUCTION

The innovation of complex technologies is now widely regarded as the major "engine" of modern economies. Complex technologies are those that cannot be understood in full detail by an individual expert sufficiently to communicate all the details across time and distance to other experts. The trend toward complexity is suggested by Figure 1, which shows that in 1970 complex technologies comprised 53% of the thirty most valuable world manufactures exports. By 1996, that number had grown to 76%.

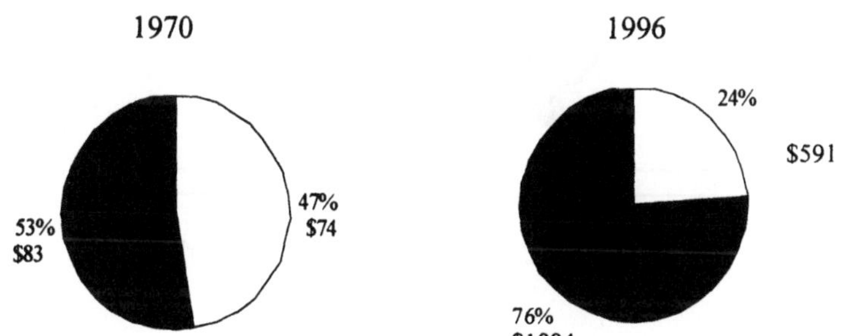

Figure 1. Thirty Most Valuable Product Exports: A Comparison for 1970 and 1996 (In $ Billions)

Understanding Complexity, Edited by Ragsdell and Wilby.
Kluwer Academic/Plenum Publishers, New York, 2001

A central point to recognize is that complexity represents a wave that is moving across nearly the whole spectrum of technologies. An illustration of this process is the ancient art of agricultural planting. One of us owns a farm in Iowa. When the farm was purchased in 1970, corn and soybeans were planted with a mechanical planter. The slotted discs in the bottom of the seed boxes were linked through gears and shafts to a wheel and when the wheel turned the seeds dropped through a tube into the row. Today planting is done with a technological system that is dazzling in its complexity. Planting density is changed meter by meter. The planter is controlled by a microprocessor and a digital database derived from data collected by the combine over the last three harvests. The planter operator has a liquid crystal display that shows every seed planted. The seeds are planted using a vacuum system and all this is guided by a global positioning system.

Two things stand out regarding the innovation of complex technologies. One is that it has become a routine activity. The second is that we do not understand how it is done. This is because the repeated innovation of complex technologies is carried out by equally complex organizational networks, like the one represented in Figure 2.

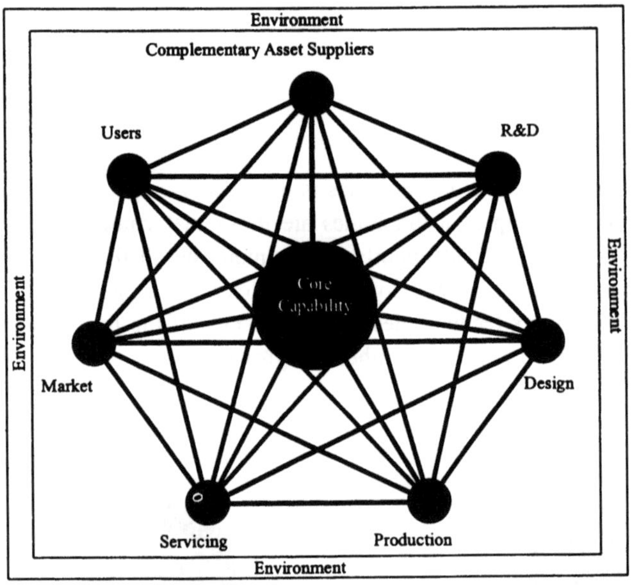

Figure 2. A Complex Innovation Network.

CASE STUDIES

We have carried out over the past five years various case studies of the evolution of six complex technologies, as listed in Figure 3. Without exception, these case studies featured networks of companies that were expert at techniques such as systems integration. Indeed, systems integration was a core capability for many of these networks. Yet when asked how they do systems integration, the answers from engineers and other members of work groups inevitably took on a metaphorical character. The common picture they suggest is that they combine knowledge and capabilities and get something different than the sum of the parts.

- General Electric's Jet Engine Turbine Blade
- Sony's Compact Disc
- Hewlett-Packard's Cardio-Imaging Technology
- Intel's Microprocessor
- Sony's 3.5-Inch Micro Floppy Disk Drive
- Varian's Clinac Linear Accelerator

Figure 3. Case Studies of Complex Technological Innovation

If complex technological innovation is one of the most important forces in the world today, then it has important consequences for those who seek to forecast. Fortunately for forecasters, the vast majority of complex technological innovations are incremental; they involve enhancements of an established technological design. Forecasting the next one or two incremental advances is possible because these innovations are heavily path dependent. That is, the history of the evolution of the technological design is a good predictor of the next few increments. Also, over time a technological community will have emerged that shares a consensus regarding the next several increments in the path, or trajectory, of the technology. If you want to know what will come next, ask members of the community and you will come very close to getting a consensus.

The best-known example of this community-based technological consensus is found in the semiconductor industry. Over the last twenty years, those networks innovating semiconductors have been guided by "Moore's Law," a community-wide expectation that there would be a fourfold increase in technological capacity every three years.

PATTERNS OF INNOVATION

But what about longer-term forecasting? Our case studies have suggested that the innovation of complex technologies is characterized by three patterns. Figure 4 illustrates these patterns and how they often are linked.

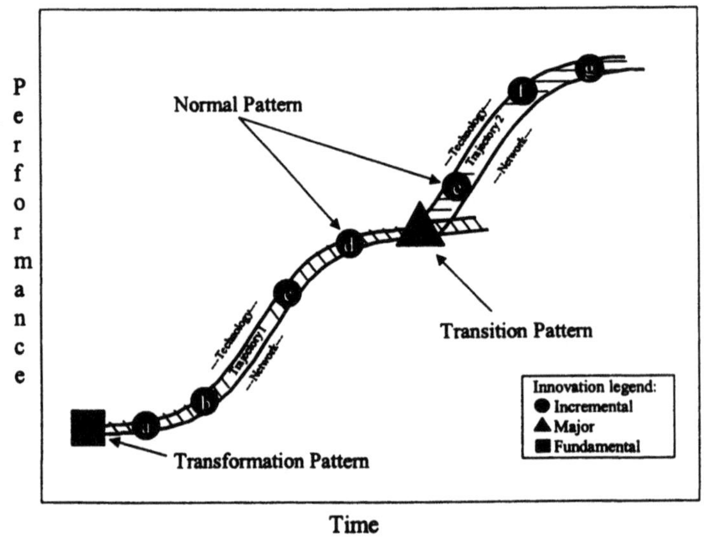

Figure 4. Three Innovation Patterns

The first pattern is indicated by the square at the left-hand side, which we label a transformation. Transformation patterns have two general characteristics. First, they involve a first of a kind technology design. Second, they involve the creation of a first of a kind organizational network. This network has core capabilities that never existed before. In our case studies, the example that best illustrated this was the creation of Varian's Clinac radiation therapy technology, which utilized a linear accelerator to create artificial radiation.

The normal innovation pattern is next, and it involves an established network enhancing an established technology design. This is nicely characterized by the last twenty years of incremental innovation on Intel's microprocessor. The greatest profits tend to accrue in the normal pattern. It is the preferred context for innovation for most networks and the one toward which they migrate.

The third innovation pattern we call a transition. The transition pattern usually requires an existing network to integrate new core capabilities, and it involves a major modification of the technology. The way transitions typically occur is for capabilities to be integrated that previously existed in another technological sector. Our best case study example of a transition was Hewlett-Packard's cardio-imaging technology. In this case, the technology evolved from using ultra-sound to the use of magnetic resonance imaging. A merger of two networks was required for the transition to take place. One network centered on core capabilities located in Philips, the Dutch electronics company, and the other on Hewlett-Packard.

Seeking insights into the adaptive characteristics of the innovation patterns, we used a framework composed of five components, displayed in Figure 5. Taken together, these components provide innovative networks with the ability to self-organize and adapt during the evolution of particular patterns or when there are pattern changes.

- Core Capabilities
- Complementary Assets
- Organizational Learning
- Path Dependence
- Selection Environment

Figure 5. Characteristics of Innovation Patterns: A Framework for Network Self-Organization

The transformation pattern is the highest risk, most chaotic, and most difficult to forecast of the three innovation patterns. Major surprises and discontinuities are commonly encountered. Elements of both the network and the technology design experience significant adaptation. Figure 6 outlines the characteristics of the transformation pattern.

As noted above, a transformation innovation involves creating new core capabilities. Those capabilities are commonly distinguished by being the product of synthesis. They involve systems integration of existing technologies and organizational capabilities with the result that something is produced that did not exist before. Complementary assets are usually composed of some knowledge and skills that were pre-existing and some that are new creations. The organizational learning associated with the transformation pattern usually involves organizations that carry out search and discovery across a broad range of knowledge areas. We call this cosmopolitan learning. Commonly the solutions to problems come from quite unexpected areas. Since transformational innovations are new, they are not significantly bounded by history. Thus, there is little path dependence.

Self-Organizing Factors	*Before Transformation Innovation*	*After Transformation Innovation*
Core Capabilities	Established (or none).	Mostly new. Some uses for existing capabilities, but highly unpredictable.
Complementary Assets	Established (or none).	Some new. Role for hybrid or fused assets.
Organizational Learning	Local or regional (or none).	Cosmopolitan. Broad expansion of search and discovery processes.
Path Dependence	Strong or medium (or none).	Weak. Past development of innovation context somewhat important.
Selection Environment	Powerful or moderately powerful (or none).	Strong institutional role; limited market role.

Figure 6. Typical Technological and Organizational Changes in the Transformation Pattern

The normal pattern is characterized by substantial continuity, as indicated by Figure 7. In the normal pattern, core capabilities are enhanced incrementally with each innovation.

There are unlikely to be any major changes. The same is true for complementary assets. Most of the learning needed to carry out innovations occurs locally (i.e., close to previous learning) within the existing network and surrounding technological community. This learning emphasizes ever-greater depth as opposed to breadth. Innovation is very path dependent and clear market signals play a strong role.

Self-Organizing Factors	*Before Normal Innovation*	*After Normal Innovation*
Core Capabilities	Established.	Established. Some enhancement by way of altered linkages to complementary assets.
Complementary Assets	Established.	Established, plus any or all of the following: *Reconfiguration of existing assets; *Upgrading of existing assets (e.g., from generic to specialized; from specialized to integration into core capabilities); *Linkages to new assets.
Organizational Learning	Local.	Local. Often more focused than prior to the increment.
Path Dependence	Strong.	Strong. Often even more strongly linked to the past than prior to the increment.
Selection Environment	Powerful market.	Powerful market. Signals more meaningful than prior to the increment.

Figure 7. Typical Technological and Organizational Changes in the Normal Pattern

Innovations that occur within the transition pattern are characterized by the changes delineated in Figure 8. It is usually necessary for some new core capabilities and complementary assets to be created as part of the transition. Organizational learning is based on search and discovery processes we call regional—broader than the local learning of the normal pattern, but not as broad as the cosmopolitan learning of transformations. Path dependence is at a moderate level, and the selection environment features a combination of market and institutional (including governmental) influence.

Self-Organizing Factors	*Before Transition Innovation*	*After Transition Innovation*
Core Capabilities	Established.	Partly new, due to some combination of: *Adapting existing capabilities by modifying approaches and routines; *Adding new capabilities (e.g., cross-sector spillovers); *Eliminating capabilities that have become outdated.
Complementary Assets	Established.	Partly new; specialized assets often added.
Organizational Learning	Local.	Regional. Broadening rather than deepening. Altered heuristics and experimentation.
Path Dependence	Strong.	Medium. Influence of past depends on whether transition arises from past events.
Selection Environment	Powerful.	Moderately powerful. Institutions become as important as markets.

Figure 8. Typical Technological and Organizational Changes in the Transition Pattern

SUMMARY

Given these pattern characteristics, what can be said of the future of forecasting the innovation of complex technologies? We have been particularly struck by the growing use of industry-wide road mapping as an element of forecasting. Road maps serve as vehicles for coordinating innovative networks and other members of the technological community during the normal pattern. Mapping exercises also assist these actors in anticipating the emergence of transitions.

Beyond road mapping, it is clear that most transitional and all transformational patterns will be difficult to anticipate. Forecasting them will require many more detailed insights into the underlying technological and organizational dynamics of learning, path dependence, etc., than are currently available. Foresight regarding transformations in complex technological innovation may be a long time in coming. In the meantime, however, we believe the ideas presented here can help minimize *oversights*.

MAKING TECHNOLOGY FORESIGHT (and SYSTEMS STUDIES?) USEFUL

Alan L. Porter

Technology Policy and Assessment Center
Georgia Tech
Atlanta, GA, USA

INTRODUCTION

"Systems analyses" rarely are taken seriously -- we rhapsodize with each other, but don't often add value to policy or managerial decision processes. "Technology foresight analyses" fare similarly. Neither form of analytical process is worth doing unless it affects decision making. What can we do to change this? I address this issue for technology foresight, with overtones pointed to the systems community.

This paper draws mainly upon two resources. First, seven of us have been drafting a paper on future directions for technology foresight analyses (Porter at al, 2000) The next section highlights how technology foresight is adapting to changing information resources, technologies, and, especially, user needs. The second resource is an ongoing project sponsored by the U.S. National Science Foundation (Porter et 1998-2000). The following section draws on that work to help understand why technology analyses are so poorly utilized by technology professionals and managers. The final section suggests ways technology foresight and, by implication, systems analyses must change to add real value.

CHANGES IN TECHNOLOGY FORESIGHT

The various forms of technology foresight analyses have evolved significantly over the past 50 years. Further significant change appears likely in substance and objectives. Technology foresight methods are evolving as well, but that topic is addressed elsewhere in this volume by one of the co-authors, Hal Linstone. A notable shift in the concerns and consumers of technology analyses is occurring -- from the former prominence of technical issues driven by military aims to innovation issues driven primarily by economic competition. Emerging technologies meriting foresight analyses are increasingly science-

Understanding Complexity, Edited by Ragsdell and Wilby.
Kluwer Academic/Plenum Publishers, New York, 2001

based, altering the premises of forecasting. Social considerations will play larger roles in future technology decision making.

"Technology foresight" encompasses:

- Technology (and Technological) Forecasting (TF)
- Technology Assessment (TA)
- Technology Roadmapping
- (national) Technology Foresight studies, and
- Competitive Technological Intelligence (CTI).

Since the inception of TF, driven mainly by U.S. military foresight needs in the 1950's, we have seen the emergence of TA in the 1970's, driven by concerns about the "unintended, indirect, and delayed" (Coates, 1976) societal effects of changing technologies. In the 1990's we've seen strong growth in national technology foresight studies and CTI, (Ashhton and Klavens, 1997) driven largely by economic competitiveness concerns.

The prime target users of American technology foresight have changed from military to civilian. Their information needs center upon how to direct technological advance for competitive advantage. There are myriad variations on this theme. For instance, a corporate unit might undertake technology roadmapping to anticipate coming generations of vital technologies and to understand how they interact. This information should guide R&D toward technologies with the highest strategic importance for the unit. Moreover, this technology roadmapping might well be combined with *product* roadmapping to ascertain how technologies fit with which business interests. And, these analyses would likely extend beyond the company's own development planning to explore *external* developments. The payoff there comes from tracking competitor prospects, identifying potential collaborations, and properly gauging opportunities and their attendant risks.

This change in information needs has strong ramifications for technology foresight utilization. The scope of analyses must encompass the economics. Whereas once TF for the U.S. military might have focused tightly on technical issues vis-a-vis superiority to the Russians' projected systems, technical issues must now be located in broader socio-economic systems considerations. "Technology Forecasting" must be recast as "Innovation Forecasting" (Watts and Porter, 1997). That implies explicit consideration of the socio-economic system and the important influences on the implementation of technological change (innovation).

Technology itself is changing. Today's information technologies change more rapidly than did their predecessor industrial technologies. Traditionally, firms in certain sectors (e.g., pharmaceuticals, chemicals) have relied heavily on science for their innovations. That tendency appears to be spreading to many other industries.

New forecasting methods are required as technology becomes more science-based. TF relies upon the relatively orderly cumulation of incremental improvements to justify methods such as trend extrapolation. However, scientific progress is less orderly because the underlying driver (research emphases) is based on self-organizing principles. Self-organizing phenomena result in sudden improvements. Forecasting science-based technology is rather like forecasting the weather. It requires rich contextual data and quick recognition of implications. For example, when we decide whether to picnic today or not, we use weather channel information to get a current regional pattern interpreted in terms of direction and speed of weather fronts to estimate the chance and severity of rain. We want succinct information; we don't want a 5-minute report rich in "interesting" information (read "noise"

– e.g., history, national weather patterns, temperatures in 18-odd cities) that doesn't relate to our immediate decision.

Note that both of these shifts in technology foresight, from military to economic competitiveness interests, and toward science-based developments, call for richer information resources. Good analyses must draw upon more contextual information. Fortunately, recent advances in electronic information resources can help fulfill this need. Various data mining approaches thus come to bear, but bring additional user acceptance problems (addressed in the following section).

Analytical methods must become more systems oriented, recognizing the complexity of modern technological innovation processes (Rycroft and Kash, 1999). Multiple interacting forces must be considered. This heightens the value in using multiple methods within a given technology foresight study (Lemons and Porter, 1992) – something long understood, but also more demanding on all concerned. In particular, this complexity in topic and methods intensifies difficulties in getting busy decision makers to spend the time to comprehend and use more intricate technology foresight analyses.

The broadening perspective on technology foresight should encourage renewed interest in the potential effects, i.e., technology assessment (TA). Interest in technological innovation extends to its institutional matrix, and its effects on corporate activity, economic structuring, work life, political processes, and international relations. Private sector involvement with "technology roadmaps" and technology-oriented strategic plans prepared by corporations and industry associations find a public sector counterpart. Technology outlooks, projections, and reports are being done by U.S. government agencies such as the Department of Energy, the Department of Transportation, and the National Institute of Standards and Technology. Other nations engage in national technology foresight studies (e.g., Japan, Germany, UK, South Korea, and many others).

Assessing the *impacts* of science-based technology (TA) is also more challenging than gauging the implications of incremental advances in established technologies. Discontinuous change can cause system upheavals whose implications are tough to figure.

We are increasingly aware of the close interrelationships of technology with its institutional settings, economic exchanges, environmental and health implications, and quality of life. Sustainability considerations further increase the complexity of technology foresight analyses to address life cycle issues. Again, the usability challenge is to make all this information accessible and directly relevant to the busy manager or professional. The danger of information overload is extreme.

It is against this background that one can evaluate the perspective of current and future users of TF. Users of technology foresight include companies, government agencies, international bodies (such as the UN and the WTO), and non-governmental organizations (NGOs). The latter can include trade unions, lobby groups, and universities. Governmental organizations use technology foresight for policy purposes. These can include guidance with regard to R&D funding, development of national critical technologies, military developments, and environmental assessments to support legislation, for example. The private sector has been a user of TF, although it has been typically restricted to larger companies. The interesting question is, however, how the spectrum of users of TF may change in the next twenty years, and also how their needs will change. Following the arguments above, one can identify the shift to broadened information requirements.

From a large company perspective, innovation increasingly depends on external collaboration on aspects ranging from research to product development to customer service. That demands more external information than in the days of greater vertical integration.

Especially for firms committed to continual product innovation (e.g., HP, 3M) this implies judicious use of CTI and technology foresight to inform technology strategy.

Small technical companies have relied on innovativeness to capture market. In the early decades of the 21st century, rapid technological change implies that they too need to be technologically informed. That has rarely been the case in the past, excepting the immediate awareness of the initial entrepreneurs. Small firms often plead lack of time and resources to invest in luxuries such as technology foresight. There is a need for development of easily comprehensible, timely, and cheap sources of technology foresight for small companies, and possibly even for particular customers. Perhaps, web sites could provide easy access to technology foresight?

Who should engage in technology foresight? I'd like "everyone" involved with technology to do so. Every applied scientist, each design engineer, product developer, production manager, technology marketing professional, and so forth should get informed on where the related technologies are likely to be heading. Every intellectual property manager, strategic planner, CxO (e.g., CIO, CTO, CEO), legislator, agency executive, and (in short, everyone with a stake in technological innovation) needs to see the value in technology foresight and use it well. The next section explores why this isn't happening.

UTILIZATION ISSUES

I have made the case why technology managers and professionals should want analyses of emerging technologies. The logic seems compelling. Yet the track record shows such analyses roundly ignored. Why?

Our research on mining external data resources for technology foresight purposes suggests difficulties regarding immediacy and credibility to busy professionals and managers (Porter et al, 1998-2000). Important considerations include:

1. Most digests of external technology developments convey "interesting" information on a wide range of aspects, but they don't directly answer the pressing questions of the current tasks engaging the target users.
2. Our studies often are provided as daunting reports – lengthy, time-draining treatises that interfere with getting the user's immediate tasks completed. Many would-be users prefer specific answers to their questions delivered "just in time."
3. Results deriving from data mining are unfamiliar to most would-be users and, hence, difficult to understand. The same holds for many technology foresight methods being less than transparent.
4. Counter intuitively, most systematic analyses are less credible than highly intuitive opinion. The complex analyses deriving from empirical measures don't offer the reassurance of having someone to blame if things go awry. ("It's not my fault; I found this great expert and followed his "best possible" advice.")
5. Studies rarely are provided in time. A survey of 26 technology professionals and managers collaborating in our NSF project found them typically needing results fast – within a day (21%), within a week (45%), or within a month (24%). Well-tailored foresight studies choke on such time constraints. Off-the-shelf technology foresight products may be available, but aren't likely to be neatly targeted on the issues being decided (see #1).

6. The form of technology foresight studies may put off busy users as too demanding (e.g., grappling with complex models or learning to use interactive software to get to specifics).
7. On the one hand, technology professionals and managers don't want outsiders involved with their proprietary considerations. On the other, they don't want to take the time to learn how to perform the analyses themselves. One solution is to engage lower level staff to learn and do the analyses, but they may lack credibility. It is hard to effect successful coupling of analyst to user (Kostoff and Geisler, in press).
8. Utilization is surely enhanced by engaging the intended user throughout the analytical process. But, not only is this time demanding, it requires the user to learn unfamiliar methods. That can be intimidating.
9. A good counter to #8 is to nurture a few chosen technology professionals or managers through the process to "seed" the organization with champions for technology foresight. However, this effort can be undone too readily if that champion leaves the organizational unit – a frequent occurrence these days.

SUGGESTIONS

Remedies deriving from the NSF results(Porter at al., 1998-2000) include:

A. Challenge institutional barriers to electronic information access. Ready access to information resources is vital to several of the following suggestions. However, in a nasty "chicken and egg" dilemma, it needs successes to justify the costs of obtaining access and the risks in opening up the organization to external information (see G).
B. Generate topical analyses on demand through use of scripting and information product templates to provide technology-of-choice analyses "instantly."
C. Build strong analyst –user rapport. "User training" appears an essential ingredient. The more key users who can be trained to appreciate technology foresight, the more the organizational culture can be made to value decisions based upon systematic foresight. This can provide the needed robustness to support maintaining foresight capabilities in the organization.
D. Customize report forms to meet the information needs of each user. Work individually with each target user to devise a form that provides the desired information, and only that information, in a comfortable way.
E. Packetize information – provide it in "bite-size" chunks, as needed by the decision processes.
F. Integrate analytical forms with expert opinion.
G. Document success stories for technology foresight. It only takes a small percentage of "home runs" to justify the costs of many studies. And, indeed, most studies will only contribute hard-to-document diffuse benefits. But, a few will link in readily grasped ways to major pitfalls averted or precious opportunities recognized.
H. Let natural selection take its course. If we advocates are right that exploiting information about emerging technologies provides a competitive edge, then those individuals and organizations who invest in technology foresight will outlast those who don't.

I add another set of related recommendations. The concluding chapter of Ted Gordon and Jerry Glenn's (1999) digest of factors required for successful use of futures research in decision making offers 26 recommendations. Paraphrasing the subset of these that make their way into the executive summary, futures researchers need to:

a) Assure that the target users understand "futures research" and want the results
b) Connect to the organization's strategic planning process
c) Collaborate with the decision makers in the futures research process
d) Use at least one well-understood formal method
e) Make future possibilities real for the decision makers by adding subjective descriptions to quantitative projections
f) Convey findings as stimuli for discussion, not predictions
g) Arrange workshops in which decision makers can integrate the concepts into their thinking
h) Set up formal monitoring systems
i) Provide futures analyses as part of continuing decision input mechanisms, rather than one-time studies.

It is interesting that these two lists of suggestions are largely distinct (C relates to a and c; B relates somewhat to h and i). Yet, they do not contradict each other in any regards. Testing and validation of these recommendations by others would be most valuable.

In closing, we might ask how these two sets of recommendations relate to systems analyses. Upon reflection I find this more challenging than I anticipated. I think many of the challenges are shared between technology foresight and systems analyses. However, I'm struck by the sense that "systems thinkers" may be too arrogant to even care about generating useful results. Try out recommendations B, D, and f – how many systems "true believers" would demean themselves by following these prescriptions?

The suggestion of familiarizing would-be users (c.f., recommendations C, a, and c) would seem to adapt well from technology foresight to systems analyses. General acceptance of systems principles is a necessity to initiate action upon particular analytical results.

Recommendations D, E, b, e and g suggest that systems analysts put themselves in their intended users' shoes. Develop sensitivity to those users' needs and accommodate them.

F and e suggest that modelers recognize the benefits of multiple perspectives (Linstone, 1999). Combining multiple methods could enhance credibility and acceptance of systems studies.

B, h, and i imply ongoing roles within organizations. Again, this strikes me as radical thinking for most of the systems advocates that I know.

I would welcome reactions to these recommendations by the systems community.

REFERENCES

Ashton, W.B., and Klavens, R.A., eds., 1997, *Keeping Abreast of Science and Technology*, Batelle Press, Columbus, Ohio.

Coates, J.F., 1976, Technology assessment: a tool kit, *Chemtech* 6:372-383.

Gordon, T.J., and Glenn, J.C., 1999, *Factors Required for Successful Implementation of Futures Research in Decision Making*, Army Environment Policy Institute, Atlanta GA (AEPI-IFP-0999).

Kostoff, R.N., and Geisler, E., Strategic management and implementation of textual data mining in government organizations, *Technology Assessment and Strategic Management* (in press).

Lemons, K.E., and Porter, A.L., 1992, A comparative study of impact assessment methods in developed and developing countries, *Impact Assessment Bulletin* 10:3:57-65.

Linstone, H.A., 1999, *Decision Making for Technology Executives: Using Multiple Perspectives to Improve Performance,* Artech House Inc., Norwood, MA.

Porter, A.L., Carlisle, J.P., and Watts, R.J., 1998-2000, *Mining Bibliographic Information on Emerging Technologies,* U.S. National Science Foundation, Management of Technological Innovation project (DMI-9872482).

Porter, A.L., Coates, V.T., Farooque, M., Klavans, R., Lapid, K., Linstone, H.A., and Pistorius, C., 2000, The future of technology foresight, presented at the International Symposium on Forecasting, Lisbon.

Rycroft, R.W., and Kash, D.E., 1999, *The Complexity Challenge: Technological Innovation for the 21st Century,* Pinter, London.

Watts, R.J., and Porter, A.L., 1997, Innovation forecasting, *Technological Forecasting and Social Change* 56:25-47.

Section Six

Y2K

WHY THE CENTURY DATE CHANGE OCCURRED SO SMOOTHLY

Stuart A. Umpleby

Research Program in Social and Organizational Learning
The George Washington University
Washington, DC 20052 USA

INTRODUCTION

Although nearly all programmers knew that using two digits to represent years would cause difficulties in the year 2000, repairs in most cases were postponed until the late 1990s. If equipment had not been repaired, world trade and the global economy would have been seriously disrupted. Many people might have died as a result of chemical or nuclear spills or the failure of urban infrastructure. Once the seriousness of the problem was understood, governments, corporations, and associations around the world acted to repair equipment and to prepare for possible disruptions. Although repairs could have been made in the normal process of upgrades, because repairs were not made until time was extremely short, correcting the year 2000 computer problem became the largest technical project in human history. It was the largest management challenge since World War II and the largest example of peacetime cooperation in history. Because the year 2000 computer problem became a multi-disciplinary threat and required a multi-disciplinary solution, it provides an ideal example for illustrating principles and methods from the systems sciences (Umpleby, 1999).

People who had been following the subject were genuinely surprised that the century date change had occurred so smoothly. This paper will answer several questions: What actually happened on January 1? Was there really a problem? Why was there so little disruption? Why were there no problems in countries that seemed to be less prepared, such as Italy? Why were there concerns about Russia? Why were Americans not evacuated from foreign countries thought to be most at risk? Finally, the paper will consider what has been learned from the Y2K experience and what remains to be learned.

Understanding Complexity, Edited by Ragsdell and Wilby.
Kluwer Academic/Plenum Publishers, New York, 2001

WHAT HAPPENED ON JANUARY 1?

In late January Y2K program managers from national governments, large corporations, and international organizations came together to discuss what had and had not happened. The conference was jointly sponsored by the Center for Global Security Research at Lawrence Livermore National Laboratory in Livermore, California, and the International Institute for Strategic Studies in London, England.

On New Year's Eve a lot of equipment was shut down. One report was that electric power consumption was only one third of normal. There were some failures of electric power equipment. However, since demand was low, power could be obtained via the power network from other plants. So consumers did not experience power outages.

In order to keep the Information Coordination Center from being overwhelmed with reports of minor failures, a criterion was established for "reportable failures." However, because of quick fixes there were very few failures that met the criterion of "reportable failures."

The most affected equipment in descending order of frequency of failure were PCs, servers, mainframes, networks, the internet, security systems, and embedded systems. Among failures, 80% were considered to be insignificant; 16% caused brief service interruptions; and 4% caused significant service interruptions.

WAS THERE REALLY A PROBLEM?

Given the small number of glitches reported and the virtual absence of significant disruptions anywhere in the world, one could reasonably ask whether there was a Y2K problem. Several replies are usually given to this question. First, a large number of knowledgeable executives spent a lot of money to fix equipment. Estimates are that $500 billion was spent worldwide, that $100 billion was spent in the U.S., and that almost $10 billion was spent by the U.S. government. Second, people point to equipment that did in fact fail and note that if repairs had not been made, there would have been many more failures.

Several statements made at the conference illustrate the concerns that people had. Joe Weiss from the Electric Power Research Institute was asked by a rather annoyed member of the audience, "Why didn't you tell people that the electric power grid would work?" Weiss replied, "We did not know that it would work. We thought it would work. We were fairly confident that it would work. We had done everything we could think of to be sure that it would work. But we were not sure. We might have missed something."

A similar statement was made by John Boggs from the International Air Transport Association. He said, "We were uncertain until the last moment." Finally Ed Hillard from Hewlett-Packard/ Agilent said that HP had purchased 60 iridium telephones to be sure that they could communicate among their various field offices. Apparently the people at HP were uncertain whether the telephone system would work. Finally, command centers were set up by businesses and governments around the world. Although one could say that these command centers were set up to reassure the public or to show "due diligence" in case of law suits, those who staffed them were not certain that nothing would happen. Indeed, there were numerous problems that were, in most cases, dealt with quickly.

WHY WAS THERE SO LITTLE DISRUPTION?

Those who attended the conference were genuinely surprised by the small number of disruptions. Although some said they had not expected major disruptions, no one said that there had been more disruptions than they had expected. Many explanations were offered for why there had been so few disruptions. In some cases problems were avoided by switching to manual operation. It turned out that fewer embedded systems were vulnerable to Y2K failures than was initially feared. Furthermore, there were only six to eight manufacturers of SCADA systems, which were among the systems of greatest concern. These are load-balancing systems which are used in electric power networks, water distribution networks, gas pipelines, chemical and nuclear plants, and even steel mills.

Once embedded systems were recognized as a danger in 1997, people had to learn about them, find them, determine which ones were vulnerable, and then repair or replace the affected systems. This task was called "the world's greatest Easter egg hunt." Fortunately, those who started this task early made public what they had learned. Hence, those who started late learned from those who started early. They could look for and replace only the vulnerable equipment. This greatly simplified the task.

Although the cost of not fixing the problem was very high – loss of market share or even bankruptcy – the cost of making repairs was small. The money required to make repairs was usually less than one percent of operating revenues, and often less than half of one percent. In contrast, the cost of converting computer systems to handle the Euro is three to six times more expensive. The small sums of money involved meant that managers did not have to wait for budget approval. They could spend the money immediately and adjust budgets later. This sped up repairs. Once top management became aware of the magnitude of the danger and hence the importance of the project, top management became committed and made the necessary resources available.

Email and the Internet were widely used to increase awareness and to share technical information, and all presenters agreed that the Internet was crucial to the successful outcome. International associations were used to increase awareness, to provide technical information, and to gather reports on vulnerability and progress.

There is low information technology penetration in much of the world. So, many countries simply did not have much equipment that needed to be repaired or replaced. Repairs were thus made quickly, even though some countries started late.

For me the greatest surprise at the conference was the news that multi-national corporations had been very active very early in all the countries in which they operated. Large corporations were acting abroad as they were at home. They not only repaired their own equipment, they also worked with their critical suppliers. For example, Shell Oil Company held awareness-building seminars, conducted technical training seminars, and made available information on vulnerable systems.

There was unprecedented cooperation among all affected organizations – businesses, governments, and associations. Y2K was perceived as a common threat. Due to economic interdependencies, people realized that everyone had to be ready in order for anyone to be ready.

Critical sectors, primarily electric power and telecommunications, were addressed first. After these were fixed, attention was given to less critical equipment.

WHY WERE THERE NO PROBLEMS IN ITALY AND RUSSIA?

Italy is an example of a country where problems were anticipated. It is an industrially advanced country, which means there is vulnerable equipment, and Italy started late in its attempts to repair its systems. The country did not appoint a Y2K coordinator until late 1998. He did not have an office, staff, or budget for several months. The first government-wide meeting on Y2K was held in Italy in either August or September of 1999. Yet when the Y2K coordinator began checking with key utilities, he was told, "It's been fixed." This sequence of events created both concern and then disbelief. People first worried that Italy had learned about Y2K much too late to repair vulnerable equipment. Then they thought that it was not possible for Italy to have repaired its equipment so quickly. The explanation seems to be that the multi-national corporations had been working with the utilities and key businesses in Italy in order to be sure that they would function. However, apparently no one bothered to inform the government, which learned about the Y2K problem through the United Nations conference for country Y2K coordinators in December 1998.

In the spring and summer of 1999, there were great concerns about Russia's readiness. The issues of concern were nuclear missiles, nuclear power plants, electric power production, and natural gas supplies to Europe. Nuclear missiles cannot be launched without human action, either in the U.S. or Russia. There was no possibility that a computer failure would launch a missile. The danger was that a computerized early warning system might fail to function, and a country might launch its missiles, thinking that it was under attack. To be sure that such misunderstandings did not occur, a shared command center in Colorado was established, and the hotlines between the White House and the Kremlin, which had not been Y2K compliant, were made compliant.

Probably the greatest concerns centered around nuclear power plants and electric power generation. In the spring of 1999, it was learned that the nuclear reactor monitoring systems in Russia and Ukraine were not Y2K compliant. If not repaired, this would necessitate shutting down nuclear reactors. However, when a nuclear reactor is shut down, the reactor core continues to be hot for many months. Electricity is required to circulate cooling water. If the cooling water is not circulated, the water boils off, and the reactor melts down, similar to what happened at Chernobyl. The fossil fuel generating plants did not have the capacity both to provide power to the population and to cool the nuclear reactors. The policy in Russia was that power would be provided first to the nuclear plants and then to the population. This was certainly the correct policy in terms of preventing long-term environmental damage, but it meant that the population would be short of electricity during the winter. At the very least, this would lead to great discomfort.

An additional problem was that there were Y2K problems with the automatic systems of fossil fuel plants in Russia and Ukraine. However, the plants could be operated manually, and they were. Concerted efforts by U.S. and European organizations working with Russian and Ukrainian engineers and managers led to the repair of essential equipment. However, the complete absence of problems in these countries on January 1 was a great surprise for those who had been involved in the efforts.

THE EVACUATION DILEMMA

The U.S. Departments of State and Defense faced critical decisions in 1999. Should emergency equipment and supplies be pre-positioned overseas in order to protect American citizens living and working abroad? If supplies should be pre-positioned, where were they needed? Furthermore, should American dependents be evacuated from countries that seemed most at risk? If so, the evacuation of American citizens might lead to social disorder among the local inhabitants and hence to political problems in these countries. In the spring of 1999 U.S. government agencies realized that they needed good information in order to make these decisions.

Eventually, three sources of information were created. First, interagency working groups were formed in each country; that is, in each embassy the representatives from the Departments of State, Defense, Commerce, U.S. AID, and others shared the task of gathering information about the local utility services. Second, in addition to talking with the local providers, they also talked with people in international organizations who were familiar with the local providers of electricity, water, health care, transportation, etc. Third, the embassies questioned multi-national corporations. In the fall of 1999, there came a time when the people engaged in gathering this information realized that multi-national corporations, which were also at risk due to possible loss of services from the same utilities, did not seem to be greatly concerned. When government officials asked people in the corporations for their assessment, they were told that the corporations had been working with the local utilities, which were in good shape. Hence, by November 1999 the US government had reason to believe that there would be very few significant disruptions around the world. The decision was made not to pre-position large amounts of equipment and supplies and not to evacuate American citizens. However, in Russia and Ukraine embassy employees and their families were allowed to leave if they wanted to.

WHY DID THE GOOD NEWS NOT GET OUT?

Given that the government had information that most equipment had been fixed and that there would be few disruptions, why did this information not get out to the general public? Previously, the government had indicated that, although the electric power grid as a whole would not collapse, local disruptions of power and other utilities could be expected (Bennett, 1999; Koskinen, 1999)

When the press was asked why they did not publicize the good news, the answer was, "Good news is no news." The good news did have the effect of preventing the publication of stories about possible disasters, but it was not consider newsworthy to report that disruptions would not happen.

Other factors were at work as well. Corporations were reporting that they were ready. But it was not clear what "ready" meant. Did that mean that disruptions would not occur or that organizations were ready to deal with disruptions? Were the corporations giving an objective assessment of their situation or speaking for effect – to prevent a loss of customers or a drop in stock prices? The government had long stated that its goals were both to repair equipment and to prevent public panic. So, if officials said that the government and businesses were ready, were they giving a truthful assessment or attempting to prevent public panic? Uncertainty continued to be high, not only for the public, but also for those working on Y2K projects.

WHAT HAVE WE LEARNED?

Email and the Web were critical for the successful outcome. There was unprecedented cooperation among businesses and governments. Top management involvement expedited work. International associations acted both to share information and to gather reports on the progress of making repairs.

There were fewer problems with embedded systems, fewer virus attacks, and less "unusual behavior" (meaning public panic) than was expected. Indeed there were fewer problems for customers than usual!

Y2K led to a rise in the perceived importance of the IT sector in businesses and government. Organizations created new maps of their business processes and became more aware of their vulnerabilities to suppliers and to local utilities. Many proclaimed that the IT community rose to the challenge.

Some efforts begun by Y2K projects will continue. For example, the Department of Energy will continue to work with Russian and Ukrainian nuclear plants. The Department of Defense has become more aware of its vulnerabilities. Ambassador Percy Mangoaela, head of the UN Informatics Working Group, said that he would try to improve the functioning of UN agencies through greater use of email and the Web.

Y2K was the first crisis of a knowledge society, a crisis created by an error or oversight in man-made equipment (Muller, 2000).

Several reports have now appeared describing how organizations prepared for Y2K and what has been learned (Clarke and Murphy, 2000; McConnell, 2000; Koskinen, 2000). Although most analyses have focused on what IT managers have learned, a broader curiosity is possible. We need to have a better understanding of why people were slow to understand the seriousness of the Y2K problem (Umpleby, 1999). Y2K can be thought of as a large-scale social experiment. Societies around the world were exposed to the same threat at the same time. Consequently there is much that social scientists in many disciplines can learn from Y2K (Umpleby, 2000). It is possible to do studies which compare the response to Y2K in various countries (Pozdniakov and Umpleby, 2000). Important lessons have been learned about the management of information technology and about project management, but there is much more that social scientists and system scientists can learn from Y2K.

REFERENCES

Bennett, R., 1999, *Investigating the Year 2000 Problem: The 100 Day Report*, U.S. Senate, Washington, DC. (www.senate.gov/~Y2K/documents/100dayrpt)

Clarke, J., and Murphy, R., 2000, *The Many Silver Linings of the Year 2000 Challenges*, U.S. General Services Administration, Washington, DC.

Koskinen, J., 1999, *Fourth Summary of Assessment Information,* U.S. Government Printing Office, Washington, DC. (www.Y2K.gov/new/4thquarterly.html)

Koskinen, J., 2000, *The Journey to Y2K: Final Report of the President's Council on Year 2000 Conversion*, U.S. Government Printing Office, Washington, DC. (www.Y2K.gov)

McConnell, B., 2000, *Y2K: Starting the Century Right!* International Y2K Cooperation Center, Washington, DC. (www.iY2Kcc.org)

Muller, K., 2000, Two double binds and a double blindness: Y2K and the science system, in this volume.

Pozdniakov, V. and Umpleby, S., 2000, A Y2K point of view on economic and cultural developments in the US and Russia, in: *Cybernetics and Systems 2000*, R. Trappl, ed., Austrian Society for Cybernetic Studies, Vienna, Austria.

Umpleby, S.A., 1999, How the year 2000 computer crisis might reenergize systems science, Presented at the annual meeting of the International Society for the Systems Sciences, Asilomar, CA.

Umpleby, S.A., 2000, Some implications of the year 2000 computer crisis for academic disciplines, in: *Cybernetics and Systems 2000*, R. Trappl, ed., Austrian Society for Cybernetic Studies, Vienna, Austria.

IMPLICATIONS FOR THE SECURITY OF THE INTERNATIONAL INFORMATION INFRASTRUCTURE

Olivia Bosch

Center for Global Security Research
Lawrence Livermore National Laboratory, Livermore, CA, and
International Institute for Strategic Studies,
London, UK.

This work was performed under the auspices of the U.S. Department of Energy by the University of California Lawrence Livermore National Laboratory under contract No. W-7405-Eng-48.
This paper provides some preliminary assessments which are part of a larger project organised jointly by the Center for Global Security Research (CGSR) at Lawrence Livermore National Laboratory and the International Institute for Strategic Studies (IISS), London, on international security implications of threats and vulnerabilities to critical information infrastructure.

INTRODUCTION

This paper explores which measures used to deal with the Year 2000 (Y2K) problem are also applicable to the future security of information networks in critical infrastructure. As information and communications technology (ICT) become ever more widespread, protection of both the information and the infrastructure upon which it relies becomes important as financial transactions and economies become increasingly global and interdependent. The Y2K problem, whereby computers with two digit year formatting are not able to read 00 as the year 2000, meant that errors would occur in many processes and calculations, such as billing, manufacturing and trade, in both the public and the private sector around the world. After January 2000 had passed, other computer incidents occurred such as the distributed-denial-of-service attacks in the US in February 2000 and the ILOVEYOU virus attacks globally in May 2000. Furthermore, high-level attention to international ICT developments was paid in July 2000 at the G-8 Summit of industrialised countries and at the annual meeting of the United Nations Economic and Social Council (ECOSOC); these all sustained the importance of information infrastructure and its protection.

This paper will first assess the steps undertaken by governments, industry and international organizations to deal with the Year 2000 (Y2K) issue. The motivations behind the effort and requirements for fixing this problem varied not only between these three groupings but also within them. For example, advanced western nations and developing countries were influenced by different factors. The second part of the chapter will make some preliminary assessments of how representative the Y2K problem was of other threats to and vulnerabilities of information systems. Then, various factors that contributed to managing the Y2K issue will be examined for their applicability to information infrastructure protection. While government and industry cooperation was a predominant element in dealing with Y2K, this and other factors, such as the spread of good management practices, and the work of international organizations are equally important to critical information infrastructure protection.

The Y2K problem is presented as a case study of a threat to or vulnerability in the information networks of critical infrastructure. Given the potential large scale and global scope of the impact of the Y2K problem, it was a valuable example from which to assess the vulnerabilities of critical infrastructure, and thus also the means of protecting it. The Y2K issue was both a simple technical problem readily understood and also a complex issue in that it involved all sectors of society, including those critical infrastructures with complex interdependencies which, if they were to fail, had potential for large-scale adverse consequences.

The critical infrastructure of a nation is defined as those assets upon which the security and economic and social viability of a country rely. In this paper, they comprise the assets of the energy and communications sectors upon which financial, governmental and emergency services depend. Other sector assets such as food, water- treatment plants, manufacturing plants, the oil and gas industries, and transportation are included in broader definitions of critical infrastructure, but these sectors, too, rely ultimately on energy and communications systems. Over the past decade, information technologies have increasingly been used in these infrastructure assets to improve controls and monitoring of processes as well as business transactions, and therefore the security of these information networks in the critical infrastructures require examination.

The Y2K problem was expected to impact upon the energy and communication sectors in different ways: on the one hand with an immediate noticeable failure, and on the other a cascading or perpetuation of errors in systems that otherwise appear to be working as normal. These two sectors and error types, while general and broadly defined, indicate that critical infrastructure is not a monolithic entity, but composed of different technologies and related systems which require protection of not only information but also physical assets.

MOTIVATIONS FOR DEALING WITH Y2K

Governments, industry, and international organizations are the focus of examination as they own, manage and regulate critical infrastructure, and these three types of entity each differed in their motivation and approach to fixing the Y2K problem. The focus on these categories, however, does not mean the efforts of many other individuals and organizations that also took an active role in dealing with the issue were unimportant. Critical national infrastructure underpins the operations of governments and economies, therefore, its protection is also a matter of national security.

Governments were motivated to deal with the Y2K issue by concern with security and law and order, as well as the smooth running of government services. The confidence of the public therefore was a prerequisite and led governments to encourage disclosure of information about how the problem had been dealt with. Such a view was adopted primarily by the US and UK governments which led the world in efforts to raise awareness of the problem. The US and to a lesser extent the UK and Japan are heavily dependent on computer systems and having the largest financial centers as well as strong industrial and IT sectors, these countries had global interests in making sure the problem had been fixed. Reflecting part of this global concern, the US and the UK were the main contributors to the World Bank's Information for Development Program (known as infoDev) which offered non-OECD countries grants for assistance with their Y2K problems. In developing countries, which are less dependent on computerized or digital automated systems, raising awareness of the problem in a similar way to that adopted by the US was believed to be unnecessarily alarming to their publics, or not considered as important enough in light of other more pressing social, political, and economic problems. This view was held in October 1999,[1] even after the June 1999 UN-mandated 2nd Global National Y2K Coordinators conference in New York, where more than 170 countries attended to discuss and compare how countries were preparing for the impact of the Y2K problem.

Not surprisingly, security was the main motivation behind action by ministries of defense and other national security organizations. They too relied on effective communications and energy supplies to be able to maintain existing military operations, and assist with humanitarian relief operations to alleviate Y2K-induced disasters, if required. Additionally, much public attention was devoted to the possible impact of the Y2K problem on nuclear-armed missiles, and thus the nuclear weapons states, particularly the US and Russia, turned attention to showing the public that these particular weapon systems and associated early-warning systems were secure.

While public attention was on high profile areas such as nuclear-armed missiles, aircraft, and civil nuclear power plants, these are complex systems which are potentially vulnerable to hazardous impacts of all kinds, and thus have been heavily regulated for safety and made subject to stringent physical protection measures for decades. Redundancy of safety mechanisms, the shift from manual to automated systems and layered contingency procedures were ways to reduce the risk and impact of errors in such complex systems.[2]

Industry was motivated to fix the Y2K problem by concerns about liability for damage caused by Y2K disruptions and by the impact upon their profits, as well as their reputation. In the US and UK, and in varying degrees throughout Western Europe, the private sector owns and/or operates upwards of 90% of national critical infrastructure. In developing countries, many electricity utilities and communications facilities are still owned and operated by government, though deregulation of these sectors has already begun. The approaches to fixing the Y2K problem by advanced countries, therefore, were quite different

[1] During the Asian workshop of the CGSR-IISS Y2K Seoul Conference, October 17-18, 1999, discussion among participants from southeast and east Asia indicated a range of approaches regarding publicity of Y2K: from reporters finding their own stories, to government telling the media what the news was, to no publicity which unnecessarily alarmed a population that was not heavily dependent computer systems.

[2] See, for example, Nancy Leveson, *Safeware: System Safety and Computers* (Menlo Park, CA: Addison-Wesley, 1995); and Ronald L. Enfield, "The Limits of Software Reliability", *Technology Review*, Vol. 90, No. 3 (April 1987), pp. 36-43.

from those undertaken by developing countries. Not until after January 1, 2000 did factors emerge that more clearly explained the low level of Y2K disruptions in developing countries. These factors included the degree to which a relatively small number of vendors of critical infrastructure could provide a more uniform approach on how best to conduct fixes in their respective sector. Secondly, developing countries "leap-frogged"[3] by benefiting from the research and "homework" done in more computer-dependent countries, in particular from the best practices that were spread by multinational companies abroad. Lower levels of automation and government-owned infrastructures meant that centralised approaches facilitated Y2K repairs as well as contingency planning once the urgency was understood.

The major international governmental organizations associated with dealing with the Y2K problem were primarily the United Nations-mandated International Y2K Cooperation Centre (IYCC), and those organizations that have a regulatory role which includes assuring safety of a sector. These organizations include the International Telecommunications Union (ITU), the International Atomic Energy Agency (IAEA), the International Civil Aviation Organization (ICAO), as well as international trade and industry associations such as the International Energy Association (oil and gas) and the World Association of Nuclear Operators (WANO). The Electric Power Research Institute (EPRI), the Information Technology Association of America (ITAA) and the American Petroleum Institute, while specific to the United States, also had international influence on fixing the problem which was a potential safety hazard if not fixed. The international governmental organizations issued statements and guidelines and requested that their member states take appropriate measures to ensure that Y2K was fixed or its effects mitigated. Though repair or replace efforts were paid for by either member states or private owners of equipment, and any expected disruptions would have local impact, the international organizations were able to use their regulatory role to place pressure on and coordinate governmental efforts in these particular sectors.

International organizations were able to harmonize efforts among governments worldwide, while large multinational companies were able to spread best practices; the private sector had unprecedented cooperation with governments in advanced countries while multinational corporations worked with governments in the developing countries in which they operated. While it was not formal, there was a system of interaction between governments, industry and international organizations to deal with repair efforts. This cooperation also continued when global Internet schemes were being established to monitor and share information about potential Y2K disturbances as they occurred in each time zone over the roll-over period from December 31, 1999 to January 3, 2000.

EXPECTATIONS IN 1999 OF POTENTIAL IMPACT OF Y2K

From late 1998 to early 1999, before widespread disclosure of information about how effectively efforts to fix the Y2K had been, the expectations among the public and some organizations as to how the Y2K problem would impact upon them varied from apocalyptic disruption to not much more than a nuisance. The US, by far the most reliable on ICT, appeared to be the only country in which a small sector of the population, for example,

---

[3] Presentation by Bruce McConnell, Director, International Y2K Cooperation Centre (IYCC), for CGSR-IISS conference "International Security Aspects of the Year 2000 Issue: Preliminary Assessments of "What Really Happened" and Lessons To Be Learned", January 24-25, 2000, at LLNL at http://cgsr.llnl.gov/Y2KLessonsAgenda.html; and *Y2K: Starting the Century Right*, Final Report of the IYCC, February 2000, at http://www.iy2kcc.org.

some survivalist and particular conservative Christian groups, expected apocalyptic or worse case scenarios; a "Christian Y2K Relocation" site was on offer in Minnesota for $64,000.[4] More of the US population, though not expecting disaster, nevertheless prepared for a heavy three-day storm, and acquired back-up or emergency generators and foodstuffs.

Various surveys were undertaken to assess public expectation and industry expectations of the disruptions that might occur. One series of surveys was conducted by Bruce Webster (Webster 1999) for the Washington D.C. Year 2000 Group, in March and May 1998, and in June and November 1999. 2000 email addressees were asked to predict the impact of Y2K in the US in five different sectors on an escalating scale of 0 (celebratory) to 10 (apocalyptic). The June survey was more optimistic than the ones in 1998. From the usable data of 337 respondents from a range of professions, there were three areas of consensus: "that the social impact will be modest (65% saw it in the 0-4 range); that the government is not as prepared as it claims to be (74% saw it in the 4-10 range) and that infrastructure...will be impacted more broadly than is generally asserted (65% saw it in the 4-10 range). Generally there was also a division between those who expected not much to occur and those who "described organizations far behind in their Y2K efforts (and usually covering it up)". By the December 1999 survey, the results had become more optimistic, reflecting a decreasing concern in the US that the Y2K problem would result in serious disruption.

While the above survey was to assess perceptions of what would occur, surveys were also made to try to assess what countries were actually doing, on the basis that as more information was known then there would be less apprehension among the public and industry. One of the most prominent studies for assessing country readiness around the globe was that conducted by the Gartner Group consultancy organization. Its surveys from autumn 1998 placed countries in rank order of Y2K readiness, based on a number of factors including the state of preparedness for dealing with the Y2K problem and on a country's dependence on information technology (IT). Warburg Dillon Read, an organization similar to the Gartner Group, used these two rankings, subtracting the readiness ranking from that of the one on IT dependency, to calculate an aggregate measure of vulnerability. Their results indicated that the countries likely to be most affected were the Philippines, Russia, China and Indonesia, with the countries least likely to be affected being the US, Netherlands and Sweden.[5] The information was based on first quarter 1999 data, and it would not have been appropriate to use it as a basis for making conclusions later in the year when developing countries as well as smaller industries had taken big steps to deal with the problem. Information that was more than three months old was soon outdated. Nevertheless, in October 1999, adequate information about Y2K readiness appears still not to have been known or believed with reports in the US indicating that Russia, and to a lesser extent, China, were still main areas of concern. Gathering information around the world on the status of Y2K efforts required travelling abroad rather than merely relying on the media whose reporters either were not technical experts or were more interested in looking for sensationalist stories.

The focus on readiness, however, was only a partial approach to dealing with the Y2K problem. While knowledge that Y2K repairs or upgrades had taken place might lessen public jitters about disruptions on December 31, 2000, by spring 1999 companies and governments

---

[4] "Taking a Hard Line On the Year 2000", *International Herald Tribune*, 12 October 1998, p. 3.

[5] "Millennium Woes", *The Economist*, August 21, 1999, p. 90.

realised that not all computers could be fixed. As importantly, governments and industries realized that it was no longer enough to have one's own systems repaired, but also to know about the efforts undertaken by the company's suppliers and distributors as they, too, either directly or indirectly, affected company operations. In particular, the interdependencies between the telecommunications and energy sectors became evident. Decisions therefore had to be made as to which computers were considered most important, or mission critical, to be fixed first for core government or business operations to be maintained. In addition to such "triage", governments and businesses, including infrastructure operators, focused on contingency planning for the rest of 1999 to ensure provision of service in case disruptions occurred. In the more computer dependent countries, a tension arose for senior officials who had to demonstrate readiness while also having to make contingency planning – the latter being less well understood by the public and, for some persons, calling into question government claims of readiness.

Even in the weeks before December 31, 1999, there remained uncertainty as to how Y2K would affect information infrastructure. The telecommunications networks expected congestion shortly after each time zone reached midnight, but this was a problem of capacity, not of Y2K. The basic switching of telecommunications networks did not have the Y2K problem to begin with, but the billing, payment, monitoring and other auxiliary functions that used the basic switching systems were expected to incur errors. There would not be a complete outage, but systems would appear to be functioning as normal except that errors occurred in the dates or in calculations using dates. There were expectations that embedded chips in energy utilities, gas and oil plants and large complex systems would, if not fixed, cause systems to stop. Embedded chips, of which only about 3-5% of the 25 billion produced had a date function, and even then only those which had data being input required assessing. The chip micro-code was almost impossible to test for every possible logic sequence and combined with the software and hardware in these large plants often being bespoke and updated according to local requirements, these factors meant there would be a degree of randomness in the occurrence of a Y2K disruption. Precautionary measures were part of contingency planning and thus many complex plants, utilities, mills, chemical plants and other factories, had controlled shutdowns or rescheduled maintenance programs. This process was made easier as the holiday season was a period of slowed activity in any event(Matthews, 1999). Additionally, as safety is a concern for many complex systems, existing in-house emergency services and contingency plans for most mishaps and disasters, for example, using back-up generators for loss of electricity or switching to manual systems, adapted those measures to accommodate the Y2K problem.

WHAT HAPPENED?

The Y2K problem has not resulted in long-term strategic or catastrophic security impacts or affected the stability of stock markets or economies. Despite glowing press reports of few Y2K incidents, it is evident that many Y2K disruptions had occurred but technical staff on duty, being the best and brightest, were able to make a quick fix or work around the problems and thus prevent or mitigate a reportable failure. The IYCC and the US Senate Committee on the Year 2000 Technology Problem each produced lists of Y2K disturbances occurring in more than 75 countries around the world.[6] The Y2K disruption to operations of the US reconnaissance and French Syracuse II communications satellites would have been the closest to affecting issues of traditional national security that have been observed. These did not, however, appear to result in serious security problems at that time.

There were few or no confirmed reports of major power outages or communications failures in countries. The main communications issues were problems of congestion just shortly after the rollover occurred in a country. Regarding energy, there were planned reductions of service as part of precautionary measures to prevent unexpected surges; there were reports of minor outages in parts of South America. There were about fourteen reports of Y2K glitches in civil nuclear power plants. These took place primarily in Japan, Spain, the UK, and US, and did not cause safety problems or (unplanned) loss of power production.

Governments and industries and international organizations each had their own requirements for implementing not only plans for repairs and contingency but also the reporting of Y2K incidents. Thus there were varied types of reports; for example, industry used help desks or existing emergency procedures to assess Y2K disruptions at the technical level, while governments were more concerned about the possible effects on public order. Reporting systems which made their results public tended to emphasize impact and failure, rather than cause. Information tended to be filtered to protect commercial proprietary information or for security reasons. Dealing with the temporary fixes as well as non-mission critical systems was delayed until some later date when resources would be available and media attention had diminished. Excepting concern about the leap year date, which also did not result in major incidents, latent effects since the end of February 2000 appear not to have been reported, even if they have occurred.

PROTECTION OF CRITICAL INFORMATION INFRASTRUCTURE

The Y2K problem varies in how closely it is representative of the kinds of disruptions expected to affect information infrastructure. While the Y2K problem is studied from many perspectives given its pervasiveness through all sectors of society, it can be seen as a surrogate attack on critical infrastructure.[7] The potentially large impact of the Y2K problem, if repairs had not been made, was compared to that from natural disaster or other worst-case scenarios. It was, however, a known risk with a fixed date and time and occurred in an "artificial" environment as many systems were turned off, on controlled shutdowns, or not subject to regular activity over a holiday period.

The Y2K problem was a vulnerability that is defined as a weakness or feature that exists within a computer or communications network and infrastructure. The Y2K problem emerged from a narrow specification in the formatting of dates in computers, with two rather than four digits used to represent the year. This was a response to what were at the time

---

[6] Appendix C - International Y2K Glitch Report, *Y2K: Starting the Century Right*, Final Report of the International Y2K Cooperation Centre; see also Appendix II - Summary of Y2K Glitches by Sector, in *Y2K Aftermath – Crisis Averted: Final Committee Report*, Summary of Committee Findings, United States Senate Special Committee on the Year 2000 Technology Problem, S. Prt. 106-XX, February 29, 2000, pp. 37-49.

[7] For preliminary lessons to be learned from Y2K are in Critical Infrastructure Protection: Comprehensive Strategy Can Draw on Year 2000 Experiences (GAO/AIMD-00-1, October 1999); the UK also produced an excellent report focusing more on IT infrastructure critical for government computer systems and e-commerce in *Modernising Government in Action: Realising the Benefits of Y2K*, Cabinet Office report to Parliament, Command Paper 4703 (London: HMSO, April 2000).

more serious problems of computer memory space and cost in the early development of computers during the 1950s and 1960s. This formatting, however, became a vulnerability as successive generations of many computers had not corrected it, and, in many cases, the specification could be worked around. Many software developers were not interested in dealing with this issue, as it was not an interesting or challenging problem to solve.

Threats, however, are intentions and capabilities that can use vulnerabilities for unauthorized use, theft or damage to the information or network or infrastructure. They include disgruntled employees, "hactivists", criminals, and professional "infowarriors" working either for governments or corporate entities. These are more widely known in the communications sector, and security measures such as password and access controls, and encryption are among measures often taken to protect information from malicious activity. While the Y2K problem was a vulnerability it was often called a threat to galvanize corrective action. The Y2K problem was often compared to natural disasters, which are categorized by some as threats given the large scale of damage and devastation that often ensues, but natural disasters do not have malicious intent.

CONCLUSION

The problems for critical information infrastructure are of two kinds: malicious attacks, and vulnerabilities of software, hardware and applications design and implementation. During the 1990s, the plant in the energy sector and in heavier industries have increasingly relied on information and communications technology to monitor plant control and processes, as well as to automate their business transactions. One of the main lessons for infrastructure owners, therefore, in particular within the energy sector, is the degree to which incorporating more IT requires a corresponding understanding of both the threats and vulnerabilities which will arise. Utility and plant owners and operators have for decades put into place contingency planning for many kinds of scenarios that affect their physical systems. Thus planning for contingency and for safety was the norm for critical infrastructure owners, and thus provided a sound basis upon which to prepare contingency planning for Y2K disruptions. Issues of information assurance to deal with the new IT issues, which tend to be better understood in the communications sector, therefore, will need to be integrated with more traditional practices related to physical plant and asset protection.

While threats can be managed by adopting some of the various security measures mentioned above, the Y2K experience illustrates the importance of adopting good practices for management of technology and information (see, for example, Braithwaite, 2000). Unlike the Y2K problem, which was seen as a one-off incident, management and security is a process, and while receiving less media attention than do hackers, it can be more important than preventing malicious attacks. More than 50% of computer disruptions arise from poor IT project implementation (75% of large computer projects are delayed, over budget and do not work as intended), flaws in software, improper configuration of computers and human error. Good IT management can minimize these, thereby allowing more attention to be devoted to dealing with malicious attacks.

To facilitate implementation of technology and information management various international standards exist and new ones are emerging. Prominent examples for consideration include the International Standards Organization (ISO) 9000 series for quality assurance management. Secondly, in the UK, in September 1999, the Turnbull Report of the

Institute of Chartered Accountants of England and Wales requires companies to raise risk assessment to the level of Board of Directors. This means accountability for disruptions and other irregularities would be in the company's Annual Report and dealt with by other regular review procedures. Thirdly, a management system standard for information security is the British Standards Institute standard BS 7799 which has been submitted to be an international standard in the ISO series.

The Y2K experience also demonstrated the effectiveness of government-industry cooperation to solve complex issues. This cooperation is increasingly important for critical information infrastructure protection. The private owners of infrastructure have direct experience of both vulnerabilities and threats to their assets, but as their systems are critical to a country's economic viability and social and political well-being, then governments require knowledge of incidents of a malicious nature to better fulfil their national security obligations. As attribution of the cause of a computer disruption, however, is difficult to ascertain in the short-term, then more information about both vulnerabilities and threats is required to improve information security. The surveys of the Gartner Group and the Washington D.C. Year 2000 Group were embryonic examples of how some IT information might be obtained. Some of the many international schemes to monitor and share information about Y2K disruptions over the rollover period are now being developed further to deal with other developments in IT and its security. These schemes included that of the IYCC the development of which the July 2000 UN ECOSOC substantive meeting discussed. The successor to the IYCC's global network of national Y2K coordinators might now focus on other IT issues, such as building communications infrastructure in developing countries and promoting e-commerce in others. The scheme set up to monitor Y2K disruptions in civil nuclear power plants might be considered by the International Atomic Energy Agency to deal with other types of accidents or incidents. Such information sharing enables government officials or company Directors and Chief Executive Officers to make decisions to mitigate or reduce potential destabilizing effects as a result of computer incidents, especially as IT systems are increasingly global and interconnected.

While the Y2K experience will not solve all future information security issues, it went a long way to demonstrating how effectively governments and industry worldwide can work together to deal with the large-scale and complex IT issues likely to occur as economies and trade become increasingly interdependent.

REFERENCES

Braithwaite, T., 2000, *Y2K Lessons Learned: A Guide to* Better *Information Technology Management*, John Wiley & Sons, New York.

Matthews, R.G., 1999, Factories plan to halt or scale back operations for Y2K, *Wall Street Journal*, December 20 edition.

Webster, B., 1999, The predicted impact of the year 2000 problem in the United States", The Spring 1999 Survey of the Membership of the Washington D.C. Year 2000 Group, released 10 June 1999, on internet at http://www.wdcy2k.org.

Y2K - "TEN MONTHS AFTER": TWO "DOUBLE BINDS" AND A "DOUBLE BLINDNESS"

Karl H. Müller

Department of Political Science and Sociology
Institute for Advanced Studies, Stumpergasse 56,
A-1060 Vienna, Austria

INTRODUCTION

This paper is devoted to a detailed ex post analysis of the Y2K-problem as well as to new ways of risk-assessments for contemporary knowledge societies in general. Basically, the paper tries to establish three main points. First, a large number of essential Y2K-propositions which have been put forward by the author throughout the period from 1997 to 1999, can be validated by their ex post outcomes. Most notably, core characterizations of Y2K as the first "global challenge", of a new type of "knowledge society", exhibiting new forms of "co-ordination problems" and entirely new "knowledge-based" "risk potentials", can and should be upheld despite the almost universal marginality of Y2K-induced disruptions and discontinuities. Second, two important groups of ex ante expectations on potential Y2K-effects turned out to be mostly misguided. In both cases however, the reason for the misjudgment lay in a homogeneous set of "alarmist" empirical assessments with respect to the high potential for regional Y2K-failures and their subsequent global "repercussions". Third, the scientific "Y2K-learning curve", in sharp contrast to the societal "learning curves", has followed a constant path along the "bottom line". It w ill be argued that the ex ante as well as the ex post attitude on part of the scientific system was an almost unanimous lack of interest which, by itself, counts as a new risk potential, inherent in today's "knowledge societies".

Understanding Complexity, Edited by Ragsdell and Wilby.
Kluwer Academic/Plenum Publishers, New York, 2001

DIMENSIONS OF THE Y2K-PROBLEM: TEN ASSESSMENTS EX ANTE AND EX POST

At the outset, ten basic propositions both from an ex ante and an ex post perspective can be put forward which characterize essential dimensions of Y2K-problems[1] and their impact for societal development in general before and after the roll-over date. Table 1 summarizes the central assertions which incorporate, additionally, new types of societal risk assessments beyond the Y2K-problem as well.

Table 1: Ten Basic Y2K-Assessments (ex ante and ex post)

Validated ex ante-Assessments

(1) The Y2K problem can be considered as the first major challenge of modern knowledge societies or, alternatively, of contemporary "Turing societies".[2] The challenge has been global and has run throughout all "Turing societies" around the world. Moreover, the challenge was universal and affected industrial enterprises, the service sector, utilities and infrastructure, private households or local and state administrations. Thus, Y2K should be viewed as the first universal and global coordination problem for "Turing societies".

(2) The challenge posed a new type of societal coordination problem which is characteristic for "Turing societies" and which has not been encountered in previous societal formations.

(3) The Y2K-problem resulted from an erroneous "encoding" or "embedding" of time-measurements and time-coordination into the basic architecture of "Turing societies". More specifically, Y2K was the outcome from codifying time as a relatively short "cycle" within the new machine code bases.

(4) The Y2K-failure was a self-inflicted and self-propagated "error" in the machine code. This "error" can be qualified as a typical "frame problem error" in which available knowledge components with respect to the "trivial" operations of Turing programs and Turing machines have been completely "discounted".

(5) The Y2K-challenge belongs to the class of most complex and most densely coupled socio-technological problems. It affects the machine code bases and their embedded hardware components, *i.e.*, chips throughout the socio-technical systems and infrastructures of contemporary Turing societies. In this sense, Y2K must be considered as a rare challenge across the actor networks and the knowledge bases of contemporary "Turing societies".

Non-Validated ex ante-Assessments

(6) Evaluated in terms of risk potentials, Y2K has been viewed by most empirical ex ante accounts as the first coordination problem of the type $SP_{t<t(s)} < RP_{t<t(s)}$. This inequality states, quite generally, that the available societal substitution power (SP) for the very short (one month) or short run (one year) ($t < t(s)$) is smaller than the Y2K-induced risk potential (RP) or, alternatively, the expected risk incidence. As it turned out, this inequality has to be completely reversed into $SP_{t<t(s)} > RP_{t<t(s)}$ for all points in time prior to December 31, 1999.

(7) The Y2K-failure has been qualified as potentially "central" both to the exchanges and to the transfers of actor networks and of knowledge pools within contemporary "Turing societies". Moreover, due to the shortage of time left, the potentially "central error" has been characterized as " intractable". While Y2K was an "intractable" problem, the Y2K-induced damages were sufficiently far away from becoming a potentially "central problem" for contemporary "Turing societies".

---

[1] For a survey of the ex ante literature, see *e.g.*, Kappelman 1997, Keogh 1997, Müller/Purgathofer/Vymazal 1999, Ragland 1997, Webster 1999, Yourdon/Yourdon 1999.

[2] The notion of "Turing societies" has been coined for a new type of societal formation which can be characterized by two necessary conditions. First, the knowledge base of Turing societies lies in a binary machine code which has been embedded in a suitable information and communication infrastructure. Second, the genetic code has become part and parcel of the operative biotechnological repertoire. Thus, new societal opportunities and "niches" can be directly "translated" and "transformed" into suitable genetic modifications and adaptations. (For more details, see Müller 1999a, 1999b, Müller/Purgathofer/Vymazal 1999) The "Turing societies", defined in this dual manner, gradually emerged in the 1950's and were in full operation by the 1980's and the 1990's.

Ex post Assessments of the Science-System:

(8) The scientific system in general has acted ex ante in a perfect "double bind". Since Y2K did not appear as a major societal "risk factor", it would have been necessary to deal with it on a new platform in an inter- or transdisciplinary fashion. But there was no ex ante need for a comprehensive inter- or transdisciplinary analysis since Y2K did not show up as a major "risk element" within the established disciplinary boundaries.

(9) For the ex post period, the scientific system has followed another "double-bind", this time with a seemingly better empirical justification. Since Y2K did not appear as a major societal "risk factor" within the established "disciplinary matrices" after January 1, 2000, it would have been necessary to deal with it in an inter- or transdisciplinary. But there was no ex post need for an inter- or transdisciplinary analysis since Y2K did not exhibit even minor disruptions within the "metabolism" of a globalized economy and a globalized "world society".

(10) Contemporary "Turing societies" must be considered as exceptionally weak in terms of the scientific monitoring and the scientific observation of the new knowledge infrastructures. The relevant empirical ex ante data for the international or the global Y2K-status have been collected by non-scientific bodies and turned out to be misleading in at least two core aspects, namely with respect to their potential impact assessments and with respect to their rankings and hierarchies. Ex post, the almost perfect "roll-over" as well as the almost perfect transition from February 29 to March 1, 2000 has effectively stopped any data production on Y2K induced failures both from the scientific and from the non-scientific world.

Within the next pages, these ten crucial Y2K-propositions will be further substantiated and justified. In doing so, a new type of societal risk analysis will gradually emerge which should be implemented within the next years in order to illustrate the actual risk profiles of the knowledge infrastructure and of the information and communication technology within today's "Turing societies".

FIVE VALIDATED EX ANTE-ASSESSMENTS

The starting point for the first part lies in a short re-assessment of major Y2K-propositions which can be upheld despite a spectacularly unspectacular "roll over" from December 31, 1999 to January 1, 2000 and from January 1 up to the present day.

At the outset, the first proposition is devoted to the scope and to the dimensions of the Y2K problem.[3] Y2K has had its origins in the machine codes or, alternatively, in "Turing programs"[4]. Due to the embeddedness of Turing programs in steering and electronic control processes, the Y2K challenge had been situated at the hardware level as well. Moreover, due to the high degree of diffusion of Turing programs and embedded chips across the socio-technical systems in agriculture, industry and services around the world, the Y2K problem had affected the fundamental metabolic exchanges and transformations within global market networks and other global societal network formations. (Assessment 1)

Second, Y2K must still be viewed as a new type of societal coordination problem which re-combines three separated features, namely complete predictability, a necessity for effective problem-solutions and a universal and global threat or involution potential. In a morphological manner, Table 2 highlights different groups of societal coordination problems.

---

[3] The Senate report on Y2K from February 1999 starts out with the phrase that "Y2K is the first big challenge of the information society." (Bennett and Dodd 1999)

[4] A short remark on the terminology must be made. Following the notion of "Turing societies", the concept of "Turing programs" refers to any codified material which is suitable "embededded" in a "machine code" and, in a "Turing environment", *i.e.*, in mainframes, PC's, etc.

From Table 2, the notion of "transferability in time" requires some additional comments. In general, a societal coordination problem is to be qualified as "time-transferable" if it can be delayed or reproduced in time without specific temporal boundary conditions or limits. Taking unemployment as a reference case, then a substantial unemployment reduction is one trajectory among many possible national pathways only. In principle, unemployment may persist in time indefinitely, sometimes at high levels, sometimes at lower ones, sometimes rising, sometimes falling. In this manner, consumption of heavy drugs, fatal traffic accidents, violent crimes and many other societal phenomena are to be qualified as time-transferable coordination problems since they are reproduced anew from year to year without any temporal limit imposed on their effective reduction or abolition. Consequently, Y2K has belonged to the rare occurrences of non-transferable coordination problems, having an exact and insurmountable "expiration date", namely the time interval from 23.59 p.m. on December 31, 1999 to 0.00 a.m. on January 1, 2000.

Table 2: Major Types of Societal Coordination Problems

	Predictability			
	Yes		No	
	Local	Global	Local	Global
Transferable in Time	Problem I	Problem II	Problem III	Problem IV
Non-Transferable in Time	Problem V	Problem VI	Problem VII	Problem VIII

Moreover, Y2K-solutions had to be of an effective nature, too. An "effective problem solution" is to be understood as a substantial reduction or dissolution of a specific problem. To be more concrete, an effective solution of unemployment implies a radical reduction to frictional unemployment or even a dissolution of the number of involuntarily unemployed persons. An effective solution of heavy traffic accidents lies in the radical reduction of accidents below a marginal and irreducible threshold value. In this sense, Y2K required effective problem solutions for each network actor which had to be in operation prior to a non-transferable point in time.

Likewise, non-transferable societal coordination problems with a threatening global impact had been, so far, of an unpredictable nature only. Taking fatal high-technology accidents, earthquakes, floods or other catastrophic events as "paradigmatic cases", then one recognizes immediately that in all these instances the element of non-predictability plays a significant role. An unpredictable high technology disaster like Seveso or Three Miles Island imposes a large amount of immediate and non-transferable coordination problems like rescue operations, safeguarding the social and natural environment and the like. In fact, advanced societies are equipped with a sufficiently developed protective capacities which safeguard their normal functioning in the case of minor or even medium disruptions.

Viewed in this light, Y2K still must be considered as an entirely new type of coordination problem, being totally predictable, requiring effective solutions and being non-transferable at the same time. Additionally, due to its machine code base, Y2K has posed a new coordination problem both for the actor networks and for the knowledge bases of contemporary "Turing societies" (Assessment 2)

Furthermore, the core of the Y2K-problem has consisted of a highly revealing inversion of the traditional modes of time-encodings and time-measurements. More specifically, time has been structured or, alternatively, "structurated" (Anthony Giddens) towards the end of the so-called "Piaget societies"[5] around circles of minutes (60 seconds), hours (60 minutes), days (24 hours), years (365 days) and a linear ordering of years, relying on a scale with a strange reference point (transition between - 1 B.C and + 1 A.C.) Despite this heterogeneous set of counting devices with their origins in Egyptian, Mesopotamian, Greek and Roman time cultures, the basic units of time measurement have been set, towards the transition from Piaget to Turing societies, in an exact and uniform manner. The definition for a second was "9 192 631 770 periods of the radiation corresponding to the transition between the two hyperfine levels of the ground state of the caesium-133 atom" (Barnett 1998:157), and the day has been defined "bottom up" as 86.400 atomic seconds. In this way, a sound and, above all, failure-free basis for measuring time has been established. Moreover, the introduction of radioactive clocks has enabled to determine the age of the earth itself in a sufficiently precise and consistent manner in the magnitude of 4,5 billion years (± 100 million years). In this way, a heterogeneous mix of circular-linear encodings as well as of administrative synchronizations like the global agreement on 24 time zones[6] has led to a uniform and successfully embedded "world-time" for Piaget societies. While by the end of the 1960's, time has been successfully encoded in a circular-linear fashion, the machine-code programs started to implement a relatively short two digit (99 year) and a relatively long four digit (9999 years) linear-circular version. Thus, the encoding of "real time clocks" within the Turing program bases has been undertaken both in both the long and the short version as a linear sequence of seconds/minutes/hours/days/months/years within a two digit year counter and thus a one hundred year circle (the short version with the imminent Y2K-problem) or within a four digit and, thus, ten thousand year circle (the long version with a far away y10k-problem). In both cases, a circle repeats itself indefinitely into the future. In addition, the in-built temporal machine code circle was strictly "memory-free", having no additional "counter" at its disposal for the number of circles. Thus, time differences within a single circle were recorded in the traditional and well-established ways of Piaget-societies, while time-differences between two circles posed all kinds of anomalies. The single second jump from 23:59:59 on December 31, 99 (circle I) to 00:00:00 on January 1, 00 (circle II) became the maximum time interval for this type of temporal encodings and the long interval between 00:00:00 on January 1, 00 (circle I) and 00:00:01 on January 1, 00 (circle II) was recorded as a single second jump only. It must be added that the Y2K-paradoxes with respect to time-differences were structurally similar to the "Goodman paradox" on induction, which had been generated via the introduction of new time-dependent predicates. [7] (Assessment 3)

With respect to the fourth assessment, the type of error underlying Y2K, qualified as "frame problem", requires a substantial amount of justification and commenting. "Frame

---

[5] The term "Piaget societies" refers to all societies whose essential "knowledge pool" lies in encoded or written materials like books, articles, notes on paper, diagrams, mathematical notes, etc. (For more details, see Müller 1999a, 1999b) Thus, "Piaget societies" extend over the whole range of human societies, starting with early symbolic code-producing societal formations roughly 50.000 or 100.000 years ago.

[6] From 1848 onward with the establishment of a uniform time zone for Great Britain, a homogeneous system of time-zones has been reached, following an overall agreement in 1883 which, then, has been set in practice by Canada and the United States (1883) and, following the Britain-American lead, by countries like Italy, Germany and Austria Hungary (1893), by Australia and New Zealand (1895), by India (1906), by France (1911), by the USSR (1924) and, finally, by Liberia in 1972. (On the shift towards homogeneous time-zones, see Barnett 1998:128pp.)

[7] This analogy applies, above all, to the similarities around the sensitive time-turning point. Around this point, the time-sensitive predicates lead to various confirmation paradoxes. And it is exactly around the tuning point in the 99-year circle that a large number of failures, anomalies and misleading "default options" comes into operation.

problems" (see *e.g.*, Dennett 1986, Lormand 1991) are encountered, very generally speaking, in all those instances where two "knowledge domains" K_1 and K_2 are both relevant for a decision configuration D. By a cognitive integration failure, K_1 is used for the actual decision procedure, neglecting K_2 entirely either by "forgetting" it or by "discounting" it as irrelevant. To provide a concrete example of the "forgetful" variety, suppose someone wants to go shopping either tomorrow (A_1) or the day after (A_2). One thinks about the possible advantages and disadvantages associated with these two alternatives, utilizing relevant knowledge components K_1. Finally a decision (D) is made to choose the latter alternative ($D(A_2)$). For the selection of ($D(A_2)$, another knowledge component (K_2) would have been highly relevant, namely the fact that there is a holiday in two days from now and that shops are closed. In the instance of arriving at ($D(A_2)$), however, (K_2) did not enter into the decision configuration and was "forgotten" and "left out". Quite generally, "frame problems" arise out of an insufficient or of an erroneous integration of knowledge components. As such, Y2K must still be qualified as a "frame problem" of the "conscious" type, generated through an insufficient integration of future time horizons and, more generally, of time and time-embeddedness into the present decision configuration. During the 1960's and 1970's, the "knowledge" (K_2) of a four digit change in dates in 1999 was trivially available and distributed throughout the entire community of programmers, technicians, business managers and the like. Nevertheless, immediate restrictions in computer storage capacities, cost-advantages and traditional cultural practices (K_1) generated such a large momentum that K_2 was somehow "left out" and was apparently considered as irrelevant for the space-times being. Moreover, Y2K still reveals an astonishing insight on the time horizons of human decision procedures since even knowledge with certainty about the future like the four digit change in dates in 1995, 1996, 1997 and even in 1998 could be discounted either as irrelevant or, more to the point, treated as time-transferable. Within the concrete programming settings in the 1960's and 1970's, Y2K was to be considered as a time-transferable "error" or "shortcut" whose solution, due to the triviality of the Y2K-conversion as an isolated problem, was postponed for the period close to the millennium change. (Assessment 4)

To continue the ex post validation of relevant ex ante Y2K-propositions, one can shift to Figure 1 which exhibits, following basically a taxonomy developed by Charles Perrow (1984), two basic dimensions for evaluating and for distributing socio-technical systems. According to Figure 1, one is invited to distinguish between four different clusters of socio-technological ensembles, namely between linear/loose, linear/tight, complex/loose and complex/tight systems. Moreover, each of the attributes can be scaled according to different degrees so that one is confronted with a continuum ranging from minimally loose to maximally tight on the one hand and from minimally linear to maximally complex on the other hand. (Perrow 1984:97)

Using these two dimensions of coupling and complexity, the Y2K problem still must be qualified as the most complex and most tightly coupled technology problem for a very simple reason. On the one hand, Y2K affected all possible combinations of technological coupling and complexity and extended over the whole range of very complex and tight ensembles like nuclear plants or nuclear weapons, of linear and tight socio-technological configurations like dams or continuous processing, of loose and complex units like R&D firms or multi-goal government agencies and, finally, of loose and linear assemblies like most manufacturing. (Perrow 1984:97) In this sense, Y2K has been "distributed" throughout the entire field of Figure 1. On the other hand, Y2K had a direct impact on the connections between these four possible socio-technological configurations as well. In this sense, Y2K belongs to the special or "singular" class of most complex and most tightly or densely linked socio-technological problems. (Assessment 5)

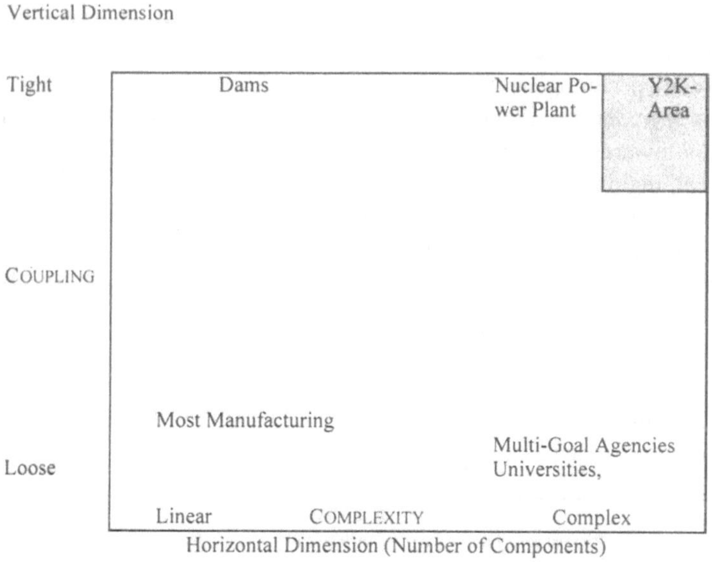

Figure 1: Two Dimensions for Socio-Technological Systems

TWO NON-VALIDATED GROUPS OF EX ANTE ASSESSMENTS

Within the present section, two groups of ex ante assessments will be presented which, due to the events after the roll-over date, have become obsolete. It should be added though, that these two expectations stood at the core of potential "Y2K-effects" and must be considered, thus, of vital importance. Nevertheless, both expected effects did not produce any tangible traces after January 1, 2000.

The first group of misleading or false ex ante propositions is linked with the scope and the dimensions of the Y2K-risk potentials within contemporary "Turing-societies". Following the available empirical assessments, it was relatively straightforward to demonstrate that problems of the Y2K-type should assume a large number of new and inverted relations between societal risk incidences, societal risk potentials and societal substitution powers. Following the second Y2K-proposition from Table 1 as well as some plausible definitions for risk incidence (RI) and substitution power (SP), one could "demonstrate" that the substitution, and thus, the coordination efforts must be operating under the following inequalities.

Traditionally, one could postulate the following inequalities for disasters in socio-technological systems (breakdowns of nuclear power plants, explosions in chemical plants, refineries, etc.), for "natural catastrophes" (earthquakes, floods, storms, etc.) and for socio-ecological disruptions (famine, epidemics, etc.).[8]

From Table 3, one obtains basically the same result which can be found in Figure 1 on different types of coordination problems already. Local catastrophes or disasters like large earthquakes, floods and the like have had a considerable local impact for (very) short, short or, although seldom, for long-term periods, but they did not exert any significant global effect on the overall performance of actor networks world-wide. In this sense, Piaget-societies which in the course over the last five hundred years had been undergoing a self-organized process towards "globalization", were exposed, under conditions of "normal accidents", to local, regional, national or restricted international crises only.[9] Towards the end of the "Piaget societies", one finds an obvious exception to the general inequalities in Table 3, namely the military build up and the military destruction potential which in the case of World War II and the new generation of nuclear weapons after 1945 has effectively transcended the inequalities of Table 3.

Table 3: Basic Inequalities for "Normal Accidents" in Piaget-Societies in the 19th and 20th Century

	Temporal Dimension Very Short Term (Days/Weeks/Month)	Short Term (< One Year)	Medium Term (< Three Years)
Local	RI > SP	RI < SP or RI > SP	RI < SP or RI > SP (seldom)
Global in Time	RI < SP	RI < SP	RI < SP

The traditional picture of regionally contained risks changed substantially, according to the ex ante Y2K-proposition, for "normal accidents" within contemporary "Turing societies". With reference to Y2K, the basic relations and inequalities in contemporary Turing societies were assumed to be of a different format.

Table 4: Basic Inequalities for Y2K-Accidents" in Turing Societies for the Period from 2000 to 2003

	Temporal Dimension Very Short Term (Days/Weeks/Month)	Short Term (< One Year)	Medium Term (< Three Years)
Local	RI > SPW or RI < SPW	RI < SPW or RI > SPW	RI < SPW or RI > SPW
Global in Time	RI > SPW	RI > SPW	RI < SPW or RI > SPW

[8] A special reference must be made that the periods of military destruction or wars should be considered as special cases in which, with World War II as the most dramatic example, even the medium global substitution power was definitely below the war-induced risk-incidences.

[9] One should add that the term "local" may comprise even regions of the size of entire nations. Nevertheless, the inequality for RI > SPW does not have a single global instance for the entire period of Piaget societies.

The first obvious inversion between Table 4 and Table 3 lay in the new relations between local and global. While Y2K might assume different patterns at the local levels, it should be qualified, by necessity, as a global challenge and a global crisis. This part of the ex ante Y2K-assesment can still be upheld "seven months after". But the other consequences, resulting from Table 4, were far off from the ex post chains of events. After January 1, 2000, the occurring risk-incidences never exceeded the available global substitution powers. According to global estimates partly from government agencies (Jaqueline Williams-Bridgers for the US State Department 1999) or from international consultants (Louis Maroccio 1999 (Gartner Group)), there should have been a substantial amount of very short-term risk incidences exactly in those areas for which only low or minimal substitution processes could take place, namely in the domain of non market networks for infrastructure and in the domain of government networks.

Moreover, the most "risky" inequality in Table 4 was the second relation in the global/short term field for the entire period of the year 2000 where the risk incidence had been assumed to be higher than the actual global substitution power. The main reasons for postulating this inequality were partly empirical, partly theoretical. From an empirical point of view, a US-State Department report had identified 88 countries with a medium or high risk potential in one of five vital infrastructure areas (energy, finance, transport, water, telecommunication).[10] The two theoretical reasons given for the global short term inequality in Table 4 are still valid ones but they depended crucially on a sufficiently adequate empirical assessment. On the one hand, the distribution of failures was assumed not to be uniform over time, but centered around a clearly recognizable peak in the vicinity of the "rollover date". This, in turn, implied that there had to be, by necessity, a peak period for which the inequality would have reached its first clearly recognizable negative "maximum". From general network theory, one could infer, additionally, that for configurations of this type "downward oscillations" should become the most likely trajectory of the overall network performance. On the other hand, a massive failure peak should bring with it negative or "risky" secondary, tertiary, quartary, ..., n-ary effects.[11] Thus, a failure in energy transmission between a utility company and a firm will have a negative secondary effect if and only if the output relations of the firm with other firms or with private customers will be hampered. Likewise, tertiary effects could be defined in a similarly recursive manner so that it was relatively clearly recognizable that Y2K-problems should have sent a large number of n-ary "shocks" throughout the global market networks for goods, services and infrastructure, the non market networks for infrastructure and, finally, to government services. All three reasons combined offered a basic justification for the global short term-inequality RI > SPW. But, due to the highly misleading ex ante data base, the market networks or the infrastructure nets both nationally and globally never experienced any types of secondary, let alone tertiary or n-ary consequences.

Moreover, the effects of Y2K-induced damages were described as a "global lottery" with a large amount of strange features. First, the participants in the global lottery were assumed to be regions.[12] Additionally, the lottery was assumed to be operative for a 12

---

[10] See, once again, William-Bridgers 1999.

[11] The term "secondary effect" had been defined as follows. Starting again from a basic network relation of the format $A \Rightarrow B$ and a non-substitutable or "non-re-establishable" failure in $A \Rightarrow B$, then a negative (positive) secondary effect lies in any relation between A and its environment or between B and its environment which is negatively (positively) affected by the failure in the $A \Rightarrow B$ relation.

[12] These regions had been defined in the following way. Since around September/October 1999, the number of the world population had surpassed the 6 billion threshold, it seemed useful to split the global "Turing society"

month period, starting on January 1, 2000 and ending on December 31, 2000. Moreover, the lottery had a peculiar distribution of gains and losses, with a very small number of "lottery gains" and a large amount of "lottery losses". The basic justification for this distribution lay, once again, in the short-term global inequality in Table 4 which implied, *inter alia*, that the overall global network performance for goods and services for the year 2000 should fall well below the expected or predicted growth values. Once again, well defined "global lotteries" had not been encountered throughout the entire period of Piaget societies and had, thus, to be qualified as a typical new feature of contemporary Turing societies, their new risk incidences and their new risk potentials. Finally, the strangest feature of the "global lottery" was assumed to lie in the fact that the outcomes of the Y2K-lottery should be linked only indirectly to the degree and to the amount of substitution efforts prior to January 1, 2000. While one could establish, on a priori grounds, a significantly positive correlation between the degree of substitution efforts and the subsequent Y2K-induced damages, the correlations were assumed to be far from perfect or even far from highly significant. Once again, it was believed that there would be a substantial number of regions with the combination "high substitution effort/low performance" and "low substitution effort/high performance". Without going into the detailed and still valid justifications for this point, the effective outcomes of the "global lottery" turned out to be significantly different to the ex ante propositions. The first "global lottery" had a unique distribution with no single looser and, thus, does not qualify as a "lottery" in any meaningful sense of the word. The interesting element though was the peculiar fact that the "all win" configuration had been reached largely independent of the amount of Y2K-preparations and Y2K-adaptions. In this residual sense, terms like "chaotic expectations" or "global time-quake" with very low values on the damage scale though, still offer an interesting and "deep" summary of the global Y2K-conversion processes. (Assessment 6)

The second group of non-valid ex ante Y2K-propositions was centered on a relatively new complex framework which could and can be applied both to the actor-networks and to the knowledge pools of contemporary Turing societies. This new multi-component framework for the "Great Transformations" within modern societies[13] was characterized by two main attributes, namely by metabolic transformations and by maintenance/repair processes. The basic results lay in a theorem which remains, quite obviously, valid irrespective of the actual Y2K-outcomes. This theorem states very generally, that all types of metabolic-repair networks (MR-networks) are faced with two different types of "network failures". On the one hand, MR-networks can be distributed in a largely independent manner which implies that the substitution capacity for small network failures is low and that the probability for a complete network collapse is close to zero. On the other hand, highly connected MR-networks exhibit high degrees of substitution capacities and, thus, a low probability for regional network failures. But densely coupled MR-networks show a peculiar feature, namely a so-called "central component" which, due to its non-re-establishable character, has the potential of "crippling" the entire network. Thus, the notion of a potentially "central error", having become "intractable" in the course of the second half of the 1990's, had been established as an additional ex ante proposition.

---

into regions of 250,000 persons respectively, dividing, thus, the global "Turing society" into 24.000 separate regions with a population of 250,000 each.

[13] On this framework, see esp. Rosen 1991, Casti 1986, 1988, 1989a,b, 1992.

Seven months after the roll-over, it is quite obvious that the danger of a Y2K-induced "central error" has been practically zero for any point in time prior to January 1, 2000. Both the linkage structure within the global MR-network and the risk-potential for major infra-structural network components did not even come close to critical "threshold values" relevant for a "central error" to occur. The actor network formations as well as the knowledge pools of contemporary Turing societies, both conceptualized as metabolism-repair configurations, were never too densely interwoven in a non-robust manner. (Assessment 7)

THREE EX POST ASSESSMENTS

The final section of this paper will focus mainly on the science system in general which has acted and reacted towards the Y2K-problem in an almost perfect "lock in-mode", being stuck within two very similar "double bind-configurations". It will become the main task in the final assessment part to stress three major "science failures" with respect to the Y2K-problems. First, the scientific inactivity prior to the roll-over date, while understandable in terms of a "double bind-configuration," had no cognitive or rational justifications and must be qualified as both an irrational and an irresponsible "high risk-strategy". Second, the scientific inactivity after the roll-over, while at first sight a seemingly rational strategy, can be characterized as questionable, to say the least. Finally, the scientific system as a whole has failed to establish sufficiently reliable mechanisms for observing and monitoring the core "knowledge based-processes" and their ICT-infrastructures. Thus, the science system was and still is confronted with a "double blindness" with respect to a data-based assessment of the Y2K-risk potentials within contemporary "Turing societies" both ex ante and ex post.

Before entering into a detailed discussion of the final three assessments, the notion of a "double bind" has to be laid out with sufficient clarity. In logical terms, a "double bind-configuration" is present whenever two conditions are fulfilled simultaneously. Starting from an initial configuration I and from a specific action or action-sequence A, one is led to a special domain D, which must be implicitly or explicitly "accepted" in order to be able to follow A (Condition 1). By "selecting" D however, a new decision has to be made which, by necessity, leads back to the initial configuration I. Aside from the well-known "catch 22" episode, typical "double bind-situations" consist in "self-destroying" imperatives like "be spontaneous", "act autonomously", "laugh freely", "withstand orders", etc. Here, the acceptance of "following an order" leads immediately back to the initial configuration, since "following an order" is incompatible with "spontaneous" or "autonomous" actions, with "free laughing", with "withstanding orders", etc.[14]

With respect to Y2K, both the ex ante and the ex post double bind configuration assume the same underlying "deep structure". The initial situation I was characterized by a "big sleep" and by a pre-dominant inactivity within the different segments of the science system. Within the established disciplinary boundaries, Y2K did not constitute a challenging scientific problem, a vital threat to the science system in general or a major societal risk factor. Thus, it would have been necessary to deal with the scientific side of the Y2K-problem by establishing new inter- or transdisciplinary platforms (D), suited for the purpose of Y2K-investigations. But any such initiative was bound to fail since Y2K did not

[14] For a more theoretical background see Watzlawick and Weakland 1990.

constitute, within the boundaries of available "disciplinary matrices", a challenging scientific problem, a vital threat to the science system in general or a major societal risk factor.

In this manner, a stable pre-roll-over "Y2K-double bind" emerged for the science system as a whole. At this point, it seems worthwhile to point out, first, that the novelty of the Y2K challenges "transcended" the existing disciplinary boundaries completely. Viewed in terms of core-competencies, no scientific discipline was equipped to deal with the complex ramifications of Y2K issues. For the domain of information or computer sciences, Y2K was a highly trivial problem to be solved by the end-users of computer hardware and software. For "science studies", Y2K was an ill-defined problem at best since it dealt mainly with programs, codes and program-errors. For technology assessments, it was uncommon to investigate the societal fate of trivial program errors. For economists, the neoclassical background offered no suitable link between model constructions and program errors in time measurement. Social scientists in general were very slow to adapt to the new features of contemporary "knowledge societies" in general and to the machine code domains in particular. A complete enumeration by disciplines would reveal very clearly that problems of the Y2K-type, due to their novelty and due to their "all-inclusive" character, were not part of the established disciplinary core competencies.

Moreover, the available data on the national as well as on the global scene were highly alarming and did not lend themselves to scientific inactivity. Surprisingly, the available international and global data published during 1998 and 1999 showed almost unanimously a bleak picture. The available data differentiated very clearly between different nation groups and between different national, regional or sectoral degrees of potential Y2K-damages. Except for an "optimistic" global financial data base, most empirical assessments gave the impression that the on-going Y2K-adaptation work was too small, too fragmented and too late for a secure roll-over process. Nevertheless, the "sounds of scientific silence" remained the only tune during the entire pre-roll over-period.

Probably the most important cognitive reason for the ex ante "double bind" lay in the apparent triviality of the Y2K-conversion problem. More concretely, the Y2K problem revealed a fascinating "mimicry" which was partly responsible for the slow societal reaction and largely accountable for the non-reaction of the scientific system. As an isolated problem of program conversion, Y2K must be qualified as highly trivial and as effectively solvable. Given a well-defined small program, using two digit year codes, it was a matter of utmost simplicity to transform the program into a four digit version. Even for large programs, Y2K modification meant a tedious and year-long search for two digit time codes. Nevertheless, the success of this search process could be effectively tested and the search and substitution processes by themselves do not constitute an exciting scientific issue at all. In this sense, the Y2K-problem appeared, at first sight, as a minimally linear and minimally loose technology issue, clearly situated in the lower left domain (loosely coupled/linear) of Figure 1. But the first assessment part has clearly shown that beyond the Y2K-mimicry Y2K-problems were not confined to isolated conversion routines but had become a highly embedded and widely distributed societal failure of time-co-ordination. Two digit codes had been used, according to proposition one, in a vast number of embedded chips for electronic control and steering. Likewise, Y2K conversion problems appeared, quite naturally, at the level of the machine codes and, thus, of the program level as well. Consequently, Y2K posed the rare occasion of a dual-level technology problem, distributed both across actor-networks and across the knowledge bases.[15] But due to the "Y2K-mimicry", there was, at least at first sight, no sufficient reason to transform Y2K-problems into a new "disciplinary matrix" or to investigate Y2K in an inter- or transdisciplinary fashion. In a slight variation to "Occam's razor", trivial problems should not be multiplied within the scientific domain beyond necessity. And Y2K seemed to be a clear instance far beyond the necessities for problem transformations. (Assessment 8)

The "ex post-double bind" turned out to be of a similar nature. Y2K did not constitute a major scientific problem, since it did not appear as a major "societal risk factor" within the established "disciplinary matrices". Moreover, due to the triviality of the Y2K-conversion-problems after January 1, 2000, it was not worthwhile to transform Y2K into a new "disciplinary matrix". Given the five valid ex ante assessments on the "nature" and on the "scope" of Y2K-problems, the ex post double-bind is impossible to sustain since Y2K constituted a 400 to 500 billion dollar challenge of the new knowledge and information societies or "Turing societies" worldwide. Moreover, given the financial resources, the communication processes at regional, national and global levels, Y2K emerged as a huge and unprecedented "learning experiment" for today's Turing societies. Finally, Y2K became a paradigmatic example for the new power and the new potential for self-organizing processes from the community levels up to the global level. Millions of Y2K-links, Y2K-websites, chats and other new forms of communication have been produced which helped to adapt to the Y2K-challenges. Thus, the ex post double-bind configuration qualifies, once again, as a major "science failure" and the scientific ex post inactivity is by itself ill-founded. A thorough ex post analysis on the basic dynamics and on the fundamental reasons for the global "success story" in societal self-organization would be a far more rational path to follow for events of the Y2K-magnitude. Nevertheless, the science system in general continues to keep its Y2K-spirits remarkably low. (Assessment 9)

Finally, the term "double blindness" refers to the peculiar situation that the data generation processes ex ante and ex post have created and still are producing almost totally "blind dates" or "purely white noise". The empirical knowledge on the global status of Y2K has been, on a familiar scale from 0 (total ignorance) to 10 (perfect information), close to the zero-region throughout the pre-roll-over and the post-roll-over period. While it must be stated very explicitly that the available published data and empirical guesses on part of non-scientific bodies were, in retrospect, highly misleading, the science system in general did not produce any reliable data or data guesses at all. It acted before and after the roll-over in "total blindness" and "ignorance". More generally, the term "double blindness" should be viewed as an accurate description of the "permanent ignorance" of the science system *vis a vis* ongoing core processes within the knowledge bases of today's "Turing societies" and *vis a vis* the risk potentials inherent in them.(Assessment 10)

In this manner, the Y2K-evaluation "seven months after" has come to its necessary final point. By now, it should be obvious that Y2K offers a tremendous learning potential for the basic dynamics of today's "Turing societies" and especially for their rapid self-

---

[15] An analogy for a similar configuration within the former "Piaget societies" may be constructed as follows. Suppose, a miracle substance, called "duront", has been invented around 1850 and has helped significantly to reduce the costs in paper production. Moreover, due to its miracle capacities, duront is utilized as a conservation ingredient in domains outside the paper production as well and becomes "embedded" in machines, machine tools, even in buildings, housings, in railroads, etc. The only disadvantage of duront lay in the peculiar fact that the miracle conservation ingredient lost its conservation capacity at a specific point in time irrespective of its prior utilization or production period. What would be the most rational way of dealing with the "duront problem", given the fact that the "expiration date" of duront is known thirty or forty years in advance? And what would be the most likely diffusion path for a product like duront?

organizing capacities. Moreover, the potential gains from an "upward" scientific "Y2K-learning curve" should far outweigh the barriers for overcoming both the existing "double binds" and the prevalent "double blindness".

REFERENCES

Barnett, J.E., 1998, *Time's Pendulum. From Sundials to Atomic Clocks. The Fascinating History of Timekeeping and How Our Discoveries Changed the World*, Harcourt Brace & Company: San Diego.

Bennett, R.F., and C.J. Dodd, 1999, eds., *Investigating the Impact of the Year 2000 Problem*. Senate Special Committee on the Year 2000 Technology Problem, Washington.

Casti, J.L., 1986, Metaphors for manufacturing: what could it be like to be a manufacturing system?", *Technological Forecasting and Social Change* 29:241–270.

Casti, J.L., 1988, Linear metabolism-repair systems, *International Journal of General Systems* 14:143–167.

Casti, J.L., 1989, *Alternate Realities. Mathematical Models of Nature and Man*, John Wiley and Sons, New York.

Casti, J.L., 1989b, (M.R.) Systems as a framework for modeling structural change in a global industry, *Journal of Social and Biological Structures* 12:17-31.

Casti, J.L., 1992, *Reality Rules*, 2 vol. Basic Books, New York.

Dennett, D.C., 1986, Cognitive wheels: the frame problem of AI, in: *Minds, Machines and Evolution. Philosophical Studies*, C. Hookway, ed., Cambridge University Press, Cambridge.

Kappelman, L.A., 1997, *Year 2000 Problem: Strategies and Solutions from the Fortune 100,* International Thomson Computer Press, Boston.

Keogh, J., 1997, *Solving the Year 2000 Problem*, Academic Press, Boston.

Lormand, E., 1991, Framing the frame problem, in: *Epistemology and Cognition*, J.A. Fetzer ed., Kluwer Academic Publishers, Dordrecht.

Maroccio, L., 1999, Year 2000 Global State of Readiness and Risks to the General Business Community, in: http://gartner5.gartnerweb.com/public/static/aboutgg/pressrel/testimony1098.html, March 26.

Müller, K.H., 1999a, *Marktentfaltung und Wissensintegration. Doppelbewegungen in der Moderne*, Campus, Frankfurt.

Müller, K.H., 1999b, Die brüchigen Zeit-Architekturen der Turing-Gesellschaften, ÖZG 3 (10):404–453.

Müller, K.H., Purgathofer, P. Vymazal, R., 1999, *Chaos 2000: Das globale Zeitbeben,* Döcker-Verlag, Wien.

Ragland, B., 1997, *The Year 2000 Problem Solver. A Five-Step Disaster Prevention Plan.* McGraw-Hill, New York.

Rosen, M.R., 1991, *Life Itself*, Columbia University Press, New York.

Webster, B., 1999, *The Y2K Survival Guide: Getting to, Getting Through and Getting Past the Year 2000 Problem*, Prentice Hall, Upper Saddle River.

Yourdon, E., and Yourdon, J., 1999, *Time Bomb 2000: What the Year 2000 Computer Crisis Means to You*, Prentice Hall, Upper Saddle River.

Watzlawick, P., and Weakland, J.H., 1990, eds., *Interaktion:Menschliche Probleme und Familientherapie*, Piper, München.

Williams-Bridgers, J.L., 1999, The Year 2000 Computer Problem: Global readiness and International Trade, Statement before the Special Committee on the Year 2000 Technology Problem, US Senate:Washington.

Section Seven

The Future of Systems Science

THE FUTURE OF THE NATURAL SYSTEMS SCIENCES

Len R. Troncale

Biological Sciences Department
Institute for Advanced Systems Studies
California State Polytechnic University
Pomona, CA 91768 USA

INTRODUCTION

This paper is the latest in a series that taken together provide an historical and taxonomic panorama of the new sciences of "system ness" or complexity. It does not try to discriminate between the three or four major domains of the systems sciences. It does not try to define the uses and abuses of, and confusion between, the terms "system," "systems analysis," "sciences of complexity," "science of chaos," "general systems theory," "system science," or "the systems sciences." Instead, it regards all of these activities and the holistic intellectual movements that feed into them, as one, as yet unsynthesized and unintegrated superspecialty. This paper assumes that there is sufficient similarity in all "mature" systems that transference of descriptive models, diagnosis of problems, prescription of remedies, and cross-application or cross-fertilization of tools and methods is not only desirable, but is in fact urgently needed. It also assumes that a significant increase of knowledge of natural systems will enable a much more mature social application of that knowledge. So, while the focus of the paper is on natural systems, it should be of use to both natural and social scientists. One special caveat is necessary. The organizers of the World Congress requested the specific title of this paper. They and the author realize that no one person can adequately capture the potential of a new field. We apologize in advance for any omissions or errors you discover.

PART I: TWENTY KEY DEVELOPMENTS AND GROWTH AREAS

The abundant number, size, potential for extensive influence, and sophistication of the areas cited below that border on, use, or contribute to the systems sciences indicates that it

Understanding Complexity, Edited by Ragsdell and Wilby.
Kluwer Academic/Plenum Publishers, New York, 2001

has a robust future. Some of the most recent developments cited are causing a revolution in the way science is carried out, perhaps even changing its methods forever. That revolution transcends reductionist science, while remaining dependent on healthy reductionist science. It is emergent from reductionist science, and, as in any true emergence, exhibits characteristics unanticipated in the original praxis.

The Next Generation Internet2 (N.G.I.): The New Organizations It Generates, and Its Demonstration Projects will Promote Systems Science

The recent appearance of a set of linked computer hardware networks complementary to the NSF-vBNS (very fast backbone network system) might be compared to the emergence of extensive new neural networks in animal evolution. As in that evolutionary case, they provide the opportunity and likelihood of the emergence of new phenomena. The basic configuration is a super network of regionally networked gigaPOP aggregation points capable of transferring data at very high speeds (OC48 or 2.4GB to OC192 or 9.6GB). This next generation Internet (NGI or Internet2) already consists of 178 connected U.S. universities, industry units, and governmental units. Nicknamed Abilene, for the town that was the site of the final connection between the Eastern and Western parts of our nation's first cross-country railroad (get it – connection across great distances), this net will achieve teraflop scale computing. It will be able to use petabyte archives of reductionist data for the first time, and it will allow emergence of large collaboratories of many researchers working together on a single project characterized by unprecedented speeds of data sharing, data generation, data storage, and data evolution. This is both a large-scale system, and a new system that will allow unprecedented research into large-scale systems.

The physicality of this network has caused the emergence of new organizations of science users. One is called UCAID (the University Corporation for Advanced Internet Development). Designed to promote software advances that make full use of the new network capability, UCAID simultaneously will be promoting the natural systems sciences. Many UCAID projects will develop and deliver new levels of "middleware." So called because they are midway up the software hierarchy between machine language and applications software, middleware will yield significantly advanced protocols for parallel programming and vector supercomputing. A related organization spawned by the NGI is the National Partnership for Advanced Computational Infrastructure (NPACI). It consists of heavy funding by the NSF + 42 partner institutions + 4 international affiliates. NPACI promotes usage of the NGI in science research based on models and simulations that require vast amounts of data. These organizations run workshops and conferences to help practitioners transfer tools and techniques more quickly and to help the hardware developers, software developers, and science users communicate across their many differences and specialty limitations.

But as a popular phrase in this group states, "Why should you care?" We should care because the future of assembly and use of very large-scale databases and the cause of natural systems simulation will be changed forever by this watershed event. The understanding of very large-scale natural systems is central to systems science. Just as computers allowed us to "see" chaos and fractals for the first time, the NGI and its practitioner organizations will enable us to DO systems science for the first time in history. Just as microcomputers allowed the rapid perception, spread, development, and use of chaos and fractals, the NGI will promote more rapid exploration of new frontiers of systems research as yet unimagined. It will give many new systems scientists the unique tools they need to explore what has been impossible to explore to date. It will allow unprecedented levels of collaboration and modeling in the systems sciences, and between the systems sciences and the conventional natural sciences. Many of the most fundamental obstacles to evolving a true science of systems may now be overcome, and its key questions (dimidium scientiae quaestio prudens), may now become tractable (the sign of a maturing science).

Mutual Impacts: Appearance of "Systems" Biology

Biology is a relatively old and traditional science. It began as an observational science, as all of the natural sciences did. While it will continue as a rigorously reductionist, empirically based enterprise, it has recently added several entirely new and robust components. Now it is a science of vast databases. Now it is a science that is increasingly integrative, as the multitude of specific facts and measurements are now ripe for synthesis into larger wholes of meaning. The results of bioresearch are also ripe for practical application. Entire industries (pharmaceuticals, biotechnology, bioengineering) now arise from applying its results. Its day-to-day function now requires very large research teams of cooperative laboratories. Now it is increasingly a science that uses modeling and simulation. The timing of this last development nicely complements the aforementioned recent surge in development of computer infrastructure. Further, the synergy of these developments enhance each other. The net result of these trends is the appearance of what might be called "systems" biology. Systems science can contribute to the development of "systems" biology, and vice versa.

Developments in the Human Genome Project: Genomics Needs Systems Science and Builds Systems Science

The millions of dollars spent by various national governments and private enterprise has resulted in a map of 99% of the human genome. Consider the size of this single data base; 3.2 billion base pairs of data for the human species alone; possibly 150,000 genes; most likely >2.5 products per gene given the ubiquity of post-translational processing. This last cited characteristic alone yields half a million gene products, most of which are as yet unknown. Beyond this, genes differ between humans in the species. Many of us possess variation beyond the general species genome. Recent research indicates there are widespread SNP's (single nucleotide polymorphisms) in the human population. Researchers have recently announced that they expect to assemble a database of 2.5 million human SNP's in a year's time. It is important when dealing with such numbers that we recognize that the single human mind is simply incapable of dealing with such massive amounts of data. If you consciously counted a number for every second of your life from conception to 70 years old, you would barely reach the number of base pairs in the basic human genome database. Dealing with billions is simply not within our instantaneous conscious ability. The simplifying processes, tools, and methods of systems science as rendered and delivered by computer networks, algorithms, and modeling will be needed to deal with this part of "systems" biology.

The above only invoked the numbers emerging from work on the human genome. Bioscience has already detailed 30 other complete genomes! All those millions upon millions of base pairs in databases must also be dealt with. The evolutionary and development consistencies and differences must be catalogued as we share many genes with organisms as humble as fruit flies and nematodes. Knowledge of these already complex organisms has already blessed us with insights into human diseases and aging. And this is just the beginning. Sequencers are approaching capability of yielding 10^6 base pairs in just 5 hours work. What took decades for large teams of workers can now be accomplished in an afternoon. The massive data now characteristic of "systems" biology is clearly just the beginning. It further proves the need for a rigorous future for systems science. It is needed now more than ever before or the potential of many such massive data sciences to serve humanity will be inhibited.

Developments in Proteomics and BioInformatics will Stimulate Systems Science and Use its Tools

The bountiful results of research into gene sequences are being matched by continued productivity in elucidating the products of genes. We now know the tertiary structures (three dimensional shapes) of more than 3,000 proteins down to a few angstroms resolution (this means down to the positions of their atoms and most important chemical groups). At this unprecedented level of protein resolution, we can begin to understand how vital human proteins work, and why they don't work in cases of some human diseases. We can model how they work to an extent that allows us to modify proteins to cure diseases or make useful products. With computer science and systems science as allies, fractal analysis of protein boundaries can be used to trace protein evolution as well as function.

Dealing with this vast amount of information, has transformed biology to an information science. It has taken a new specialty, bioinformatics, from obscurity to what is predicted to be a $1 billion industry by 2003 with the incredible annual growth rate of 30%. Yet America is currently training only a handful of bioinformatics specialists today, when we already need many thousands. We cannot even agree on whether tomorrow's bioinformatician should be trained as a biologist adding the computer skills, or trained as a computer scientist adding the knowledge of biology. This paper would argue that they should initially be trained as systems scientists adding computer and biology knowledge along the way.

Developments in Physionomics will Need Systems Science Tools

The vast gene databases create the field called genomics. The knowledge of gene products (mostly proteins, but which also include end-function RNA molecules) creates the field called proteomics. Dealing with the vastness of the databases for both creates the field called bioinformatics. Some now begin to talk about how all of these products interact to create the fundamental living system of the cell. Since this is primarily the old field of physiology, some call this new, vastly more detailed version of it, physionomics. We cited above circa 100,000 to 150,000 for the human cell alone. But it is now well known that many of these proteins exist as mixed (heterologous) multimers. This means that each entity we name as a protein actually has many different states, each with different mixes of subunit proteins. And these different mixed multimers might be only used by the cell in certain instances, or at certain times in the cell cycle. In doing their work, they enable the cell to exhibit the properties of "life." They do this by the vastly increased number of possible cross interactions. It is becoming common to see complex drawings at cell and medical meetings of networks of numerous proteins interacting and influencing each other to accomplish important cell functions. For example a recently reported simple signal transduction control pathway. As the proteins involved increase barely at all, the consequences increase very rapidly because of the many possible interactions. Interaction couples increase eight times and the number of rate constants (used to partly explain the group of interactions as a whole) increases thirty-five times. This is a virtual leap in complexity simply from recognizing network type interactions. While biology may be discovering new particular examples, there are several pre-existing lineages of work in the fields now clustered under systems science on how best to represent complex regulatory schema and how to manipulate networks.

Developments in Structural Cell Biology

Do not think that complexity is only found in physiology. Although cell structure is often perceived as static and stable, it is anything but that. The microtubule is increasingly seen as a very complex, very dynamic organelle whose mix and matching of a dozen components changes it function dramatically effecting such vital cell functions as cell division, cell motility, cell shape, and normal health of brain cells. The nuclear pore has evolved from earlier being conceived of as an empty hole in the nuclear envelope to a dynamic" structural complex" of >34 interlocked proteins called nucleoporins. Cell structure will someday be seen as dynamic and complex and free entity interactions in physiology. Again, biology will enhance systems science by its elucidation of particular systems, while systems science can help biology deal with the complexity.

Developments in Cellular Modeling and Simulation

I was convinced more than 35 years ago that the most complex system (per unit mass) known to humans would prove to be the cell. Since I was already passionately interested in both systems and biology, it was that prediction that led me to become a cell biologist. The cell is older and packs much more complexity in a much smaller space than even the brain (the often cited "most complex" system) partly because it was optimized by evolutionary natural selections for more time, at least 3.5 billion years. There were very few serious attempts to model this incredibly complex system until recently. The increasingly detailed knowledge of cellular molecular physiology and its interaction and control networks of recent times, combined with the above mentioned advances in hardware and software power have resulted in the feasibility of modern serious and rewarding cellular simulation efforts. Several attempts to model cell "in silico," such as E-CELL described recently in Science, are driven by the need to bring the vast detail now emerging under control for human use. The need to understand not just the isolated, reductionist aspect of cell structure and function, but also its irreducible network aspects is dependent on the success of such simulation efforts. The inevitable continued surge in molecular and network detail on cell structure and function demands increasingly complex cell models if science is to continue its work in understanding the basis for life. And this inevitability extends to the attendant need for a better systems science to act as a co-discovering agent. Advances in each specialty will enhance advances in the other.

Developments in the Neurosciences and in Anatomy

The rapid developments that are creating "systems" biology are not just on the cell and molecular levels; they are occurring at all levels of biology. For example on the organ level, the rapid advances in collection and analysis of data on the brain requires supercomputer infrastructure. At UCLA, one collaboratory is creating a terabyte database on brain activity derived from the vast amounts of data using NMI imagery. To get the most out of such large databases, created by very large expenditures of research funds, the data must be shared across vast distances by large numbers of researchers. Again, we argue that the design and delivery of such infrastructure is inherently a systems problem and these developments will stimulate and require a robust future for systems science. Similar databases are being created for the entire human body, and the development of the human embryo as compared to organ and organism development of other model organisms.

Much more robust models of neural unit function and simulations of neural networks are also emerging. The earliest systems consisted of step-by-step calculations on a tiny number of neurons. Now vast numbers of neurons can be used simultaneously, with several unique new algorithms for directing their summation, selection, and evolution. It is fascinating to see the commonalties between complex networks of interacting entities, whether they are molecules in a cell, neurons in a ganglion or brain, or species in an ecosystem. The output of these very different reductionist specialties feed into systems science when we focus on the similarities that transcend the particulars. Similarity and difference are simultaneously true, but at unique scales of study. Each specialty manifests the structure and function of "network ness" in different particulars.

Developments in Ecological Modeling and Simulation

Ecology/environmental biology is another area under study by an army of researchers and receiving considerable funding. Public concern for our environment has increased dramatically in recent years. Current results from this science now rival those of medicine for relevance to human survival and practical utility. Current estimates of the existence of as many as 30 million species, each composed of a billion individuals, all interacting together, clearly indicates that ecology is another science that intrinsically studies networks. Surely these numbers, and this complexity, rivals the interacting molecular components of a cell, or the interacting neurons of a brain. A specialty that yields results that inform us about networks, hierarchies, and cycling processes is informing us about systems science. Studies of the open systems configuration of an ecosystem, its far-from-equilibrium processes, its non-linear dynamic behavior, all further inform systems science, while the results of systems science should inform ecological studies.

Developments in Evolution Theory as Systems Theory

The mechanism of evolution is itself an intrinsically systems-based theory. Its emphasis on variation, diversity, feedback, environment, and emergence are coincident with systems processes. One dramatic alteration of evolution science has been its recent emphasis on molecular evolution and experimental evolution. The vast amount of data emerging from comparing gene sequences and the direct observation of molecular evolution, in action, in the test tube have added important new dimensions to the study of evolution. More direct observation of origins of life phenomena by manipulation of microenvironments has made a historical-descriptive field more accessible to falsification. The use of chaos simulations to understand the phenomenon of emergence enables quasi-tests of proposed models and mechanisms. Mature systems become mature by survival (dynamic stability) for comparatively long periods of time. These systems are evolved systems such that advances in either field contribute to understanding of the other.

New Multidisciplinary Centers at Major Universities

Several major universities are making unprecedented investments to build personnel and infrastructure that promotes exactly the comparisons across the conventional disciplines that yield systems science. They see the importance of the nexus between the above biological specialties to human advances and the potential of these fields for attracting external funding. Here is just a partial list of the size of investments initiated this last year for cross-disciplinary institutes at well-known institutions: Harvard ($ 50M); UC Berkeley ($ 100M); Caltech ($ 100M); Princeton ($ 70M); Johns Hopkins ($ 34M); Claremont Colleges-Keck ($50M). Our own, relatively small, non-R1 university has just invested $30M to build a biotechnology building that will focus on cross-disciplinary research and collaboratory partnerships.

Systems Engineering and Systems Production

Systems engineering was one of the first systems specialties to appear and remains one of the most developed. It has the largest number of recognized educational programs on a systems focus at recognized universities. Recent developments in microrobotics and nanotechnology are producing ever-tinier microsystems that exhibit ever more complex engineered behaviors. This tendency, and the promise of benefits it offers for society, puts pressure on design teams to understand and implement the common principles of systems design in ever greater detail. There is an, as yet unmet, need for a prescriptive general system template version of successful systems behaviors. Parallel developments in pharmocogenetics and combinatorial chemistry contribute to this need for a better understanding of the principles of systems design. In this latter case, a multi-billion dollar, international industry provides a flood of investment resources to stimulate practical results. This entire set of fields highlight the need for historical systems theory to meet and cross-fertilize modern systems analysis.

Sciences of Complexity: New Avenues of Research in the Natural Sciences

Due to the widespread availability of high computation personal computer workstations with gigabyte processors and gigabyte memories, because of more frequent use of massive parallel processing, science can now see relationships it could not see before. Chaos theory and chaotic processes can be explored at modest expense and in much shorter periods of times. The number of publications and regular conference series on scientific approaches to chaos, and chaos-based approaches to natural phenomena are multiplying in number, and the success of each is expanding. Software breakthroughs in Genetic Algorithms and Artificial Life Research have obliterated the conventional barriers between isolated disciplines such as biology, computer science, and mathematics creating a super-transdiscipline of complexity theory. Complexity research itself is virtually identical to systems science. Large numbers of graduate students in each of the conventional disciplines have been infected with great enthusiasm and curiosity to enter this new integrated specialty. The stated intention of the new specialty Artificial Systems Research is to apply the tools and techniques of all of the above to investigating the stability and fecundity of alternatively structured systems, *in silico*. This could lead to some of the first direct testing of systems structure and function in itself and apart from the particular manifestations of systems in nature.

Development of Earth Systems Science

Like biology, geology has evolved from a primarily descriptive and reductionist oriented discipline to include a much wider synthesis. Large-scale, biome or ecosystem-wide projects are now as respected as investigations on severely reduced isolated sub-sub-systems. Many departments are renaming themselves as "earth systems science" departments. These units are attempting to research the earth as a whole complex system. Their research relies on computerized networks of instruments that collect vast amounts of data. The data is so vast, that like medicine, astronomy, and biology, the data itself becomes a large system and the tools and techniques to analyze the data approach the level of systems modeling, simulation, and systems analysis. So the prediction here is the same for the other named natural sciences. Earth systems science will contribute information, demonstration, and understanding to an invigorated systems science, and vice versa.

Development of Ecological Economics

The network of international interactions in finance has many similarities to the network of species interactions in ecology, or the networks in the physiology of cells. Many aspects of chaotic systems are found in both. Both fields can use similar tools and techniques. Similar patterns are found in both and cross-inform or cross-fertilize each other. This recognition by growing numbers of workers has led to joint meetings of the ecology and economic communities characterized by great enthusiasm and excitement. The excitement comes from the discovery of tools and patterns in one science that is relevant to the other and the promised new developments that result from such insights. It is interesting that these two communities have not yet discovered the bounty that is available to both from the rapidly developing genomics/physionomics results in modern molecular biology. For example, a computer tool first developed to map physiological interactions in bacterial genetics is now being used by social scientists to model social interactions in Silicon Valley for the purpose of understanding the rapid economic growth of that area. Again, it is clear that systems science is about networks, in part, and any natural sciences that involve networks can learn from systems science, will promote systems science, while promoting themselves.

Selected Developments in Physics, Astronomy and Cosmology

These three fields of physical science represent the most reductionist of sciences. One might then imagine that they would have the least to contribute to or learn from systems science. I challenge that assumption. A cursory reading of Greene's best selling popular science treatise on the development of the String/Membrane Theory of particle physics indicates that the 3rd revolution in that theory was based on recognition of the duality of the fundamental equations. Before that recognition the workers were blocked from progress. But the equations unsolvable before the dual opposite or coupled equations were elucidated became solvable through knowledge of the matched pair relations. Now a reductionist scientist would be unlikely to have a concept of paired opposites in mind while working on a reduction-based phenomenon. But a systems science sensitive preparation would be suffused with many real examples of complementarity or duality as an inherent property of most mature systems, whatever their scale. That might have led to a quicker recognition of the logjam-breaking duality of the equations in this highly reduced case.

Do not think this is an isolated case of the utility of a systems sensitivity to advances in physics. Fritz Zwicky (the irascible, international figure in astronomy, formerly of Caltech) was the author, user, and advocate of General Morphology. He used the technique to predict dark matter and neutron stars. He also used it to predict a wide range of propulsion processes that became part of the birth of the Jet Propulsion Labs. At our Institute, we have an entire course on this technique and its potential uses in the natural sciences, engineering, and social systems design. Basically General Morphology is a guided, phenomenon-based exploration of potential process space completely predicated on systems science.

Overall, from observing all of these examples, we would claim that many of the most exciting, fastest growing science investigations underway today are systems-based and informed by a healthy systems science, while demanding future development of systems science.

Computer-Based Systems Conferences on the Internet

One of the first computer conferences was on general systems theory. It occurred because NSF sponsored a cluster of systems investigators nation-wide in the late 70's and early 80's to try the then new technique to see if it would lead to advances. Stuart Umpleby at George Washington University was the principle investigator, and many lessons were learned about how difficult debate on systems science is, and about the strengths and weaknesses of computer conferencing. Since then, the wisely-named Principia Cybernetica project led by Cliff Joslyn, Francis Heileiden has engaged an international cliental of both natural and social scientists interested in a wide range of systems topics and issues. The New England Center for Complex Systems has a series of active, computer discussion groups under the guidance of Yaneer bar Yam. There are many conventionally employed physicists, mathematicians, and biologists active at this site. The International Society for the Systems Sciences (ISSS) sponsors discussions on a wide range of systems topics involving evolution and emergence, duality, and social systems design. The rather sudden appearance of many independent and large website discussion groups on a diverse range of systems topics indicates both the interest in the need for synthesis in the natural sciences, and the strong potential future for systems science.

Developments in K-12 Systems Education

Historical Perspective. In the early days of work on a general theory of systems, most thought that training in systems science would only work at the doctoral and postdoctoral levels. It was thought that systems understanding required that level of detailed knowledge about some real systems before they could be fruitfully compared to discover and understand inherent similarities. Then in the 80's, ambitious educators began to offer M.S. degrees in various aspects of the systems sciences. Of course, the systems engineering B.S. already was fairly well recognized during this time. But B.S. degree offerings in systems science were not given the okay to proceed. Today, systems science is being taught for the first time at all levels. Sophisticated and demonstrably successful programs exist at the K-6 level, at middle and high school levels, at community college levels, 4-year universities, and even community activism levels, in addition to the formerly established, and multiplying graduate studies and graduate research levels. The purpose of this section is to list some examples to characterize this diversity. It is evident that a recent explosion is occurring that will contribute mightily to the future of systems science. Students using active systems modeling in K-12 mostly love the activity. Their enthusiasm will create a demand for this type of study in greater depth and higher educational levels. This firestorm of awareness and interest will occur just in time. We will need a very healthy pipeline of trainees sophisticated in systems science to solve modern complex systems problems. Here are some examples of the new developments in systems education.

STELLA Systems Modeling K-12. Jay Forrester of MIT long ago developed Systems Dynamics modeling on computers for use in studying complex systems problems. It made a sensation both positive and negative in the Limits of Growth studies of the seventies. The tools and techniques in this body of work continue to develop in the Systems Dynamics professional society and publications. But few know that a very well funded and organized set of projects exist to bring systems modeling to K-12 teachers and school districts. Teams of students create simple feedback models using STELLA on topics in all of the sciences, social science, and especially environmental problems. For example, the Waters Foundation sponsors a STELLA systems dynamics project that involves 10 states, 15 school districts, and is guided by 5 paid professional coordinators. The national effort in K-12 runs a biannual meeting attended by 250 K-12 teachers, principals, and school district administrators.

Systems Biology K-12. Lee Hood, co-inventor of the instrument responsible for the fast sequencing capability that made the Human Genome project possible, is the P.I. for a "systems" biology set of projects sponsored by the NSF. It is a 5-year, million-dollar project that presents biology as an information-based systems science. It uses master teachers paired with practising systems biologists in the state of Washington. It teaches the systems approach to students as young as 4 years old. The project will result in content and materials for teachers, and contributions to the development of teachers. It is a huge project involving 66 schools, 1,400 teachers, and 23,000 students.

NPACI K-12 Systems Projects. The aforementioned National Partnership for Advanced Computational Infrastructure sponsors a series of projects that makes large-scale (read "systems") modeling possible on the K-12 levels. One of its subsystems is the Education, Outreach, and Training (EOT-NPACI) unit. The EOT sponsors the GirlTECH projects for young women and MDVirtual on the high school level at the Ohio Supercomputing Center. There are at present at least 17 projects evolving from the efforts of two teams. All involve secondary education learning and technique development in areas fundamental to systems science.

The Creative Learning Exchange. The CLE is a web-based clearinghouse for systems STELLA models for K-12. It is a great resource for teachers with hundreds of simple to relatively sophisticated models involving all of the sciences and maths, and even cross-disciplinary phenomena. One can obtain models in both paper and electronic form with their explanations and pedagogical suggestions. These models are not refereed. Those models from the aforementioned Waters Foundation project are refereed.

New Approaches to Systems Education at the College and Post-Doctoral Levels

Integrated Science (Systems Science) General Education Program. This NSF-funded super project has attracted 14 grants, and >$1M for development of extensive distance learning courseware development. The yearlong course will ultimately deliver 250 rigorous cases studies of phenomena from all 7 natural sciences. It fulfills all the general education science requirements for any non-science student at any university. It uses advanced, highly interactive multimedia that has a whopping 27 built-in learning features designed into every module. These learning features have resulted in 75% of science-phobic students earning an A or B grade based on 900 challenge points per quarter. The ISGE program is billed as a "stealth" systems science program because it uses dozens of fundamental systems processes as the integrating themes that tie the 250 case studies together. So it teaches all seven natural sciences at the same time it teaches a great deal of systems science. The ISGE is targeted for rapid dissemination to the entire CSU system of 425,000 students and to many other colleges and universities nationwide. It has immense potential for reaching a vast number of students.

NECSI (New England Complex Systems Institute) Collaborative Project. This is another NSF sponsored, multi-institutional, multi-regional that plans to coordinate diverse systems education projects on the K-16 levels. The project involves many investigators who are associated with or familiar with recent developments in the sciences of complexity (systems sciences).

Sante Fe Institute Summer Workshop Series. This new series serves fewer workers but is noted for its rigor. It consists of both well-connected and highly visible conventional natural scientists and new graduate students who want to add systems understanding and tools to their conventional training.

ISGE K-12 and S.I.S. Alliance and SYSML Project. This new project would raise the level of individual projects to a self-sustaining social institution connected by the "nerve" complex of computer networking using NGInternet2. SIS stands for Systems Integrated Science. It would link all ISGE distanced learning groups across the CSU, SUNY, CUNY and other national universities. One of its main foci would be teacher training to get the multiplier effect that comes from such efforts. The need for teacher training is in the news. LA County alone needs 200,000 credentialed teachers next year. Who will supply this need? The CSU system has great potential for serving this need, and spreading systems science understanding at the same time through multiple adoption of the aforementioned ISGE. Each CSU campus has >1,000 teacher trainees in any one year. The CSU alone produced 20,000 new teachers in '96-'97. Plans for ISGE include linkage of all on-site programs to a master service unit using two innovative new computerized assessment tools that enable "Seamless" and "Evolutionary" assessment simultaneously.

Lee Hood will also design social institutional self-sufficiency into his Institute for Systems Biology NSF project. The result of several such projects could be an impressive increase in flow of students hungry for systems science throughout the educational system, and emerging into the economy. A healthy systems education will create a healthier systems science.

PART II. SIGNIFICANT NEEDS AND UNMET CHALLENGES

It would be irresponsible to only cite examples that enhance the future of the systems sciences without an honest citation of the forces that might inhibit that idealistic view of its future.

The individual depth and yet wide range of the 20 recent developments just cited are intended to prove that there is a significant explosion of mutually supporting events that will enhance the future development of systems science. But they are not the whole story. There are a number of obstacles that must be overcome to secure that rosy future. Some of these obstacles have persisted over the last four decades. They are anything but trivial. Without a clear understanding and appreciation of these obstacles, effective responses may not be forthcoming, and the otherwise healthy future potential of the systems sciences will not be realized.

Systems Education is BOTH a Promising Development and an Unmet Challenge

The last two key developments are both driving forces that will enhance the future of systems science AND potential obstacles. How can this be? This author has worked for thirty years in systems education, both on national and international levels. The history of systems education can be best characterized by initiation of new programs by enthusiastic and dedicated personnel followed by the unexpected and often unjustified dissolution of the new program in a rather short period of time. Here are some examples: the Systems Institute at University of Louisville, the Masters degree program in systems at San Jose State University, the College of Systems Science at University of Denver, the education program of the Dept. of Systems Science, part of a larger Institute at USC, and many others. Only two pure systems-based Ph.D. programs have survived (at SUNY, Binghamton and Portland State University) to the author's knowledge. And only one of these has survived the retirement of the original "star" founder.

However, the most recent new programs described below have some features that past attempts did not have. They start earlier in the educational system. They are organized and sanctioned by a wider range of experienced systems scientists, not by a sole founder. And perhaps the most important new feature, they are also supported by a range of otherwise conventional, and widely respected natural scientists. New funding sources have appeared. New applications for graduates have appeared. More recognition for the need for such training is evident in industry and government. There is also more recognition on the part of students entering the educational system that this is a good career track to select. These new features have significantly increased the chances of success of systems education programs of the future and so constitute another recent development that bodes well for the future of systems science.

Systems education is one of these formidable obstacles. Any new development in science or engineering depends upon an adequate flow of highly trained, motivated, as well as highly rewarded population of practitioners. If the systems sciences can contribute to and learn from the many key scientific developments just cited, and if it is so important to designing better complex human systems in the immediate future, doesn't our nation need a healthy pipeline of graduates in this field? Do we have in place a "pipeline supply" infrastructure to produce sufficient numbers of adequately trained systems scientists? Do we agree on the standards and curriculum for this new field? Do we have exemplars of model educational programs in place? Do we have academic infrastructures tested and proven to produce systems scientists? The honest answer to all of these questions must be "no" despite 30 years of attempts at answering all of these needs.

The Need for Key Distinctions and Discriminations

Why we should care about key discriminations? For systems education to be successful, we will need a more robust Systems Knowledge Base. But there is still much confusion and lack of consensus in the new field. Much of the misguided research, lack of communication, and lack of consensus results from a few key distinctions that are not widely recognized. Some of these include a confusion between the following differences: Systems Theory v. Methodology v. Application; Physical v. Biological/Natural v. Social Systems; confusion between Level of abstraction /deabstraction rules; Inter- v. Multi- v. Cross- v. Transdisciplinary approaches; differences between distinct Classes of Emergence. For example, on the last item, workers often confuse emergence of new scale of entities with emergence of a new level of specialization within a scale. Both the original general systems theory and the new sciences of complexity groups are prolific and promiscuous in their use of these terms. Often the field is guilty of mistaking the tool in hand for essentials of problem. "To a man with a hammer, every problem looks like a nail." It is not the purpose of this paper to try to explain how these key distinctions might be resolved, only to point them out and note that they will inhibit an effective systems education and future for systems science.

In past papers I have suggested the term "Discinyms" for "Disciplinary synonyms" (even though I believe forming neologisms is a disease of systems science). Discinyms are a source of immense confusion at the heart of systems science. There are Synonymic "Discinyms" such as cases where biology says "homeostasis"; and we say dynamic equilibrium, or biology says "autocatalysis;" and we say self-organisation or autopoiesis. Or consider for example the truly different uses of a conventional term like "sequence" in systems situations in molecular biology, organism biology, and geology. There are also Antonymic "Discinyms" such as the different uses and abuses of the words entropy and information across the disciplines. Whatever the case, these differences arise from different levels of abstraction and historical precedent in recognizing the self-same systems processes in different particular systems manifestations. Lack of recognition of the problem of discinyms has a deep impact on students trying to learn systems science, and creates immense confusion between different conventionally trained scientists when they try to talk with each other. The simple recognition of their existence could help overcome this obstacles and that is why I advocate naming the problem.

The Need for More Emphasis and Focus on Integration/Synthesis

Comparison of real systems at very deep process-levels is fundamental to systems science. That is an act of integration and synthesis. Many Nobel Prizes and the greatest revolutions in natural science are the result of deep, systems-level integrations and syntheses. It is a most highly valued human product!! But here is a critical question? Who teaches Integration and Synthesis in our schools? At what level is it taught? K-6? Middle or High School? College? Graduate? Where in society? Which Institution(s)? The embarrassing answer is that it is the most secret, misunderstood and under taught procedure in our school systems. We desperately need a Toolbox for Integration. How could we have gone 40 years without identifying and accomplishing this fundamental task? And who will accomplish it now to enable the future of systems science, and as a result enhance human futures.

The Need for Parameterization of Systems Research

In a recent conversation with Cal-Tech's President David Baltimore, he was asked 'what is the main problem that causes natural scientists to avoid systems science?' He suggested that it was the absence of adequate parameterization in systems research. It is parameterization that enables the experiments that become the "selection" agent that enables gradual improvement of theories. Some systems scientists have recognized this need in systems science. Historically the work of Miller, Odum, and Cowan has identified this critical need. For example, Miller suggested many "cross-level hypotheses" for investigation to build systems science. Forrester claims systems dynamics modeling is experimental. Workers in the new field of the sciences of complexity seem to assume that their simulation attempts are true systems experimentation. But the question is really open. Is simulation really parameterized? Craford prize-winning Odum's emergy modeling certainly has many parameters involved and tests for validity. But is it testing system ness per se or only the model of a particular system. Clearly such central phenomena as emergence, endlessly discussed by systems types, needs a more empirical approach to resolve the infinite disputes words engender. The new, suggested field of artificial systems research would directly approach this question. Until then, systems science may generally not find acceptance by natural scientists.

The Need for Alliances or Confederations of Systems Institutions

Integration of institutions is as critical as the synthesis of ideas. The last forty years of systems science is characterized by quasi-isolated systems knowledge communities. Worse than that, many of the communities exhibit considerable competitive behavior that inhibits the necessary transfer and cross-fertilization of knowledge. Consider the opinions that the following groups have concerning each other: Forrester's Systems Dynamics Soc.; NECSI-Sante Fe Institute axis; ISSS and Spin-Offs; Systems Societies by Continent; or by Systems Domain. So there are many obstacles to Alliance and Confederation much less social integration. Too often this is due to the very thing that enabled some of the knowledge communities; the presence of a super-guru or organizer ego that attracted many to the field, but then becomes overly territorial and competitive. Consider how different this is from the natural sciences. We have tried to establish confederations before, for example, the International Federation for Systems Research, or this recent World Congress. Perhaps this is a sine qua non for the future of systems science to be a healthy future.

The Need for New Methods of Empirical Refinement

Because of the wide diversity of approaches to systems science, one of the greatest obstacles to its future is the lack of a shared, internal methodology. The strongest sciences have the most specified internal methodology, as was exemplified by the Science article of J.R. Platt entitled "Strong Inference" decades ago. Without the powerful selection among various approaches and results, there simply is no progress. Because of its nature as a non-reductionist field, systems science may have to opt for empirical refinement, not empirical elimination of weaker formulations. But some selection must occur according to a consensus shared by its diverse workers. Otherwise how far can only theoretical comparisons go? There is a need for very explicit criteria on what constitutes the general, shared processes of systems to image and rigorously guide research. As a young worker in, and later Vice President and Managing Director of the ISSS, I tried to approach the task of assembling a list of such criteria to submit to a consensus. As early as 1974, in the Gen. Sys. Bulletin we presented and polled members on 33 "diagnostic questions" to be judged using 8 "performance objectives" for general theories of systems. Later, an editorial presented a list of possible criteria entitled "What would a General Theory look like if I bumped into it?" Despite the weak humor, none of these attempts resulted in what is needed...a concerted and continuous effort to come to agreement about such criteria. The future of systems science depends on such an effort.

The Need for Systems Exemplars of Research and Application

One of the toughest questions faced by systems science is, "Where Is The Value Added?" Many systems scientists I know seem to assume that all manner of human and social systems would be improved by simply applying simple systems principles. The fact that we do not have a good consensus on just what those principles are does not seem to bother them. Beyond our role as apologists for the new field, we must demonstrate to outsiders, using their values, the superiority of this approach, not assume its superiority. We need a widely recognized list of successful examples and robust transfers between disciplines. These are becoming more possible on the basis of the 20 developments cited earlier. But they are not yet widely recognized. One way that systems science can aid science lies in the Concept of "augmented hypothesis formation" of the type hinted at in our description of the third revolution in String Theory. Another is in the use of systems derived tools and techniques to handle the massive databases emerging from modern, conventional science research.

The Need for a Complete System of Systems Processes: The LPTM Case Study

It is ironic that the field that investigates "system ness" has not systematized its results. Many systems workers concentrate on only a small number of the processes that are true across particular systems. By leaving out many others because of preference or lack of study, they take the system ness out of the system they study. Some may research only the Zipf/Pareto pattern, or mechanisms of feedback, or hierarchical clustering to the exclusion of each other, or of other systems processes. There is a deep need for systems workers to become aware of the blinders they use, and of the widest possible set of mechanisms that need incorporation and investigation. It is quite understandable to delimit and make feasible by reduction the number of things you investigate, but it is not useful to do so in a way that eliminates important processes that impinge on the process you focus on in order to understand a system or system ness.

The L.P.T.M. is an acronym for a multi-year project at our Institute for Advanced Systems Studies. It is a system of 82 Systems Processes (or Patterns) (or Isomorphies) that define what is known from a wide range of literature on how systems work. It is an overall picture of the "mechanism" of "system ness." The 80+ isomorphies are the same for a wide range of natural, mature, systems when observed at a sufficient level of abstraction. They are what are true of very different particular manifestations of systems. The isomorphies form a self-organizing, mutually reinforcing set. They are highly specific, traceable, referenced, and testable. Only actual demonstrated processes are allowed. They are a highly integrated set because they are connected by >100 "Linkage Propositions" with the anticipated set of demonstrated linkages to be a much larger set. Each linkage proposition (LP) denotes a well-studied or hypothesized specific influence of one isomorphy on another. The LP's are also highly traceable, referenced, and testable. The set of LP's enable a new formal logic, >debate, and seeking of consensus. Their specificity enables seeing pathologies/prescriptions when a system isn't working. The net result of the LPTM is a "system of systems processes" that is much more easily communicated, traced, and tested because it exists as a tool on computers and can be shared by many workers.

PART III. IMPORTANT NEW RESOURCE OPPORTUNITIES

New Funding Programs at NSF, NIH, ONR

When Dr. Rita Colwell was named as the new Director of NSF, she implemented a new emphasis on Biocomplexity projects in particular, and complexity science in general. These affect not only the burgeoning genomic and molecular areas, but also environmental and ecological biodiversity. The new initiative involves several cooperating "sections" and literally $$millions. "Biocomplexity" is closely related to, could learn from, and contribute to systems science. The NSF has made a very conscious decision to promote this research and sponsor conferences in the area. As regards the National Institutes of Health: Many fertile systems science branches of research are focused on biomedical problems. Complexity will eventually extend to our concept of many costly human diseases. Many dementia's and autoimmune diseases, as well as diseases of aging, are due to genetic polymorphisms. Federal institutions with a long record of sponsoring research, have become more focused on the promised of systems science related problems. ONR has a new multi$M nanotechnology initiative. Agencies on both state and federal level are looking at environmental problems from a systems perspective. So the modern systems researcher has a much wider range of institutions to apply to for support of research programs.

New Funding Programs: Private Foundations

Another key source of new funding for systems science could be private foundations. We have already mentioned million dollar funding of systems education with powerful results by the Waters Foundation. The Gates Foundation is reported to be considering a $100M request from the new Institute for Systems Biology that would boost development in that particular approach to systems studies. Historically, risk-taking foundations have been critically important in the development of unconventional new specialties. In the forties and fifties, it was largely through private foundation efforts that the new specialty of molecular biology received its initial boost. Modern private foundations would do well to consider systems science when looking for exceptional opportunities to make a mark on future progress. Systems science has the obligation on its part to be ready for these new opportunities in funding by better defining its knowledge base, its function in society, and its standards. The new field has to establish exemplary research programs that are rigorous and prove that empirical refinement is possible. We next describe three case studies that attempt to provide such examples.

PART IV: CASE STUDIES THAT ILLUSTRATE THE FUTURE OF SYSTEMS SCIENCE

The projects of our Institute for Advanced Systems Studies at Cal Poly Pomona are aimed at overcoming some of these obstacles inhibiting the progress of systems science. While too early to judge whether they will be successful, they can be used to illustrate the types of projects that might contribute to a healthy future for complex systems studies.

Case Study II: The New Field of "Systems Allometry"

Systems allometry applies the well-established techniques and results of biological and engineering-based allometry to natural systems as systems, not as particular manifest systems. Rather than compare regularities across different species or different engineered objects, systems allometry compares regularities across completely different "scales of magnitude" of systems. Because it is based entirely on experimentally derived data published in refereed specialty journals from physics to astronomy to biology, it answers the pressing need for a method of empirical refinement and parameterization in systems science outlined above. Because it is based entirely on scientific data, it is coupled tightly to well accepted sciences. But its results could not be anticipated by any one of the established individual sciences, because its results are based on comparisons across several sciences simultaneously. Systems allometry uses established statistical tools to demonstrate highly significant log-log relationships that remain constant across widely separated specific systems from the biological to the physical. It suggests that the "rules" or "constraints" on systems design remain constant despite billions of years of difference in origin times, and despite the differences in local subsystem interactions that build any particular system. The results of systems allometry lead to startling conclusions. We see individual entitles and differences between living and non-living systems. Nature only sees scalar classes. Reality keeps spinning off the same general system from 13 billion years ago until now, but from different parts and at different levels of magnitude. The same system appears; just in different manifestations. Systems allometry reveals the deepest heart of system ness.

Case Study III: The New Field of "Artificial Systems" Research (ASR)

ASR was inspired by artificial life (AL), complex systems research carried out *in silico*. But instead of using agents based on the characteristics of genetics and life processes, ASR is based on the LPTM cited above. So it uses resources and selection pressures that are unique from artificial life research. Our Institute plans to use the very detailed systems of systems processes in the LPTM to select for improved systems design and function *in silico*. We plan to systematically remove singular systems processes from the total set and observe the consequences to systems stability and function. For example, LPTM *in silico* as ASR would allow us to directly observe the consequences of altering feedback delays or of uncoupling coupled positive and negative feedbacks, or of introducing hierarchical clustering to a set of subsystems. This would enable us to vastly increase the number of variants tested in compact computer time speeding up observable systems evolution. So ASR uses an inquiry strategy of very large #'s, trials, and time similar to AL, but in a different domain for different purposes. This set of research projects addresses the need for a quasi-empirical refinement method in systems research and for direct testing of systems hypotheses.

Case Study IV: XML-SYSML on the INTERNET

Recently HTML has been modified as XML and used to specify organized data on the computer for such well established and rapidly developing fields as chemistry and genomics data. So another Institute project would result in a single, highly interconnected set of data, available to everyone on the Internet that synthesizes all aspects of systems science. It would include linkages between systems literature, investigators, institutions, hypotheses, the LPTM, and schools of thought. It would be a cross-section of consensus on the knowledge base of systems science. Its purpose would be to encourage/enable greater, more widespread systems integration. We describe it as the ISGE "Connection" Beehive on Internet resulting from a series of NSF grants. It is planned as an open-source suite of synthesis tools and highly select, highly interconnected data with an extensively developed graphics front-end for ease of use. The SML-SYSML connection hive would serve as the manifest neural network for an evolving systems information community with defined functions. Such a resource would help form consensus in the field, would help in systems education programs, and provide a better, more complete and rapid introduction to field for newcomers. This would help overcome several of the aforementioned obstacles blocking the development of systems science.

PART V: VISION OF THE ULTIMATE UTILITY OF SYSTEMS SCIENCE

What should we conclude from juxtaposing the 20 key developments that suggest a healthy future for systems science and the several major obstacles that reduce that probability? Perhaps the real answer lies not in trying to trace a trajectory as these approaches do, but rather jump directly to the endpoint. Perhaps we should focus on what we imagine to be the absolutely necessary function a robust systems science would serve in the future of humanity.

A "Medical" Version of the Systems Sciences

There is an important and immediately recognizable precedent. There are many parallels between the history of medicine and systems science. Both began descriptively; both involve investigation of highly complex systems; both have the highest purpose of eventually leading to diagnosis of poorly functioning systems with the hope of prescription and prognosis. The words and intents of medicine can be usefully superimposed on systems science to good and instructive effect. It is important that systems science do what medicine did. In the face of daunting complexity, focus on the parts of the system that are malfunctioning and meticulously tease out how the normal system works from its dysfunctional state. Only then do you have a "handle" on, or a way through the maze of complexity of the normal system. This points out the need for a serious and detailed classification of systems pathologies. In a complex system, it is much easier to study what goes bad to tease out otherwise obscure interrelationships. Then following the medical model we could attempt to move systems science from its current fumbling descriptive phase to a prescriptive phase. The parallel even works in systems education. In medical schools the operative procedure is "watch one, do one." That is what we need for systems science. Unfortunately, our current status is more like the earlier medical practice of bleeding sick patients to rid them of imaginary bad blood. Thus, a dictum I have often cited at systems meetings; we need a Systems Hippocratic Oath: "DO NO HARM."

Systems Science is a Necessity for a Space-Faring Species

From the above potential, I predict that systems science will ultimately become a major Hallmark of the 21st Century to future historians. As I have attempted to show, it is critical to success of other hallmarks, the genome revolution, the medical revolution, and many branches of our current technological revolution. It will be even more important to the necessity that Homo sapiens become a space-faring species. We are privileged to be alive at the time of first discovery of planet systems; at least two-dozen cases are now proven. It is clear that our potential species lifetime will exceed that of the capacity of our planet ecosystem to sustain us. It is necessary that we colonize the local galaxy. Already there are serious studies of the exponential spread that could occur in this millennium. Consider how rapidly we will encounter totally unforeseen ecologies and threats. We will need a new way of more rapidly knowing and understanding what we encounter. Our survival as a species will depend on that new methodology. Reductionism will always be needed, but it is slow compared to the "augmented" hypotheses and transfer of knowledge across differences made possible by a mature systems science. So a healthy future for systems science is a sine qua non for a healthy future for our species.

SYSTEMS THINKING AND THE SOCIAL SCIENCES

Mike C. Jackson

University of Hull Business School
Cottingham Road
Hull, HU6 7RX, U.K.

INTRODUCTION

Len Troncale (in this volume) has outlined some of the important developments taking place in the natural sciences and the relevance of systems thinking to those developments. The social sciences have lagged behind the physical and biological sciences. It is for this reason that the idea of applying systems ideas to the social sciences was first introduced. When people tried to manage social systems they came across a variety of problems which they were not able to tackle using the normal scientific method (Checkland, 1981). Complexity was one, the fact that the problems they were dealing with were real world problems was another and the ability of human beings to falsify predictions about themselves, the nature of human beings, was a third. I suppose there was a feeling that systems ideas in some way could help with these problems that managers of social systems came up against.

There was a view that holism could help to deal with complexity. Instead of taking the system apart you looked at it as a whole and you concentrated on the interconnections between the parts as well as the parts themselves. To tackle real-world problems, it was felt that the notion of modelling that had come to the fore in areas such as operations research was important and perhaps could be taken over to the social sciences and used there as well. In terms of the human being, the subject of social systems, there was a feeling that social systems had been previously reduced and studied as machines or biological systems rather than as purposeful systems with their own parts and serving wider systems. Perhaps if reductionism was avoided and we treated social systems at the right level then it might be possible to overcome some of the problems that arose with using the natural scientific tradition to address people.

So we have a state, in the social sciences and in management, where systems thinking remains a hope for the future but it has, perhaps, not developed as strongly and successfully

Understanding Complexity, Edited by Ragsdell and Wilby.
Kluwer Academic/Plenum Publishers, New York, 2001

there as was outlined by Len Troncale with regard to the physical and biological sciences. We need to look for the reasons for that and probably need to be more modest in our ambitions. However, I think there are many reasons for hope and I believe that the way forward for the management of social systems does lie in a systems direction that I wish to outline.

PAST SYSTEMS THINKING

I was initially a historian and I would like to go back a little bit to look at where some of our systems ideas came from. This is my way of approaching the relationship between systems thinking and the social sciences and my way of trying to move forward the successful application of systems ideas to social and management problems.

Let's start by outlining the rich tradition of thought on which we draw, the frameworks of ideas on which we draw, and look at how systems ideas originated. We need to consider a number of different disciplines. In philosophy it started with Aristotle and Plato. In the Greek philosophical tradition systems ideas were highly prized. Later, Kant, Hegel, Spinoza, and Marx all had holistic aspects to their philosophies. Outside that main stream philosophical tradition some soft systems thinkers have drawn upon hermeneutics and phenomenology. Soft systems thinkers in the United States tradition, such as Ackoff and Churchman, have drawn on the relativistic pragmatism of Pierce and Singer.

Biology has contributed a huge amount to systems thinking. The organismic biologists, Smuts, Woodger and others saw off the mechanists and vitalists and brought a systems perspective to biology which led up to the work of von Bertalanffy, generally regarded, with Wiener, as the founding father of systems thinking. The biological tradition remains strong in systems thinking to this day. The work on autopoiesis by Maturana and Varela is a more recent contribution from biology which people believe could become a general system theory of value in domains other than biology.

From the philosophical and biological strands of systems thinking we got the birth of general system theory, with von Bertalanffy, and the inspiration for the foundation of this society. In 1954, von Bertalanffy, Gerard, and Rapoport came together to form a Society for General System Research. The impact of general system theory on sociology, management and organisation theory was vast; and in the 1950s, 1960s, and 1970s sociology, management and organisation theory were largely dominated by the systems tradition of work. We think of Parsons' work in sociology, much influenced by general system theory and by cybernetics. The contingency theory and the socio-technical systems theory traditions in management and organisation theory were dominated by the notion of the organisation as an organism. But it would be wrong to believe that sociology and management theory had to wait for general system theory. There were systems traditions of work in these two areas prior to general system theory. The mechanistic analogy looking at society as a system in equilibrium developed from the work of Parato. In Durkheim's work society was seen as an organism; and interest in the environment, and what functional imperatives had to be met by subsystems to ensure survival, developed. This thinking transferred over to management and organisation theory in the work of Selznick, and Katz and Kahn, sometimes mixed with general system theory.

Another side of systems thinking was fed by notions coming from control engineering and from systems engineering. The ideas of black box, negative feedback and variety developed in the work of Wiener and Ashby. Eventually Beer extended these ideas to the field of management and founded organisational cybernetics. The social sciences have been able to draw upon this tradition as well.

The physical sciences were largely written off in the early days as being unable to contribute to systems thinking because of their reductionism. It is becoming increasingly difficult to take that stance and it is necessary to be clear as to why that is the case. Some of the new sciences, quantum theory, the theory of dissipative structures in chemistry, chaos and complexity theory, are now having a significant impact upon the social sciences. I guess it is not since the days of von Bertalanffy that we have had a candidate for a general system theory as strong as chaos and complexity theory. With works such as Wheatley's (1991) on 'leadership', Capra's (1996) on 'the web of life' and Stacey's (1996) on 'management', we are getting chaos and complexity theory applied to social systems. There are real doubts in my mind about the ability to transfer analogies from the natural and biological sciences to the social sciences. The isomorphisms have to be demonstrated much more strongly than they have been to date. However, we cannot deny that some find them inspirational.

There is a rich storehouse of work upon which we can draw in the history of systems thinking to seek to improve our running of social systems. Len Troncale (2000) has talked about some very significant problems that are being addressed in biology and in the physical sciences, using systems thinking, which can help to improve the human condition. We all know that in organisations and societies there are other equally significant problems that require progress in the social sciences: Problems of organising the organisations and societies in which we live, problems of poverty, pollution, inequality, and under-development. These are equally significant problems and solutions have to come from the social sciences, and I would hope from social sciences that have been improved by systems ideas.

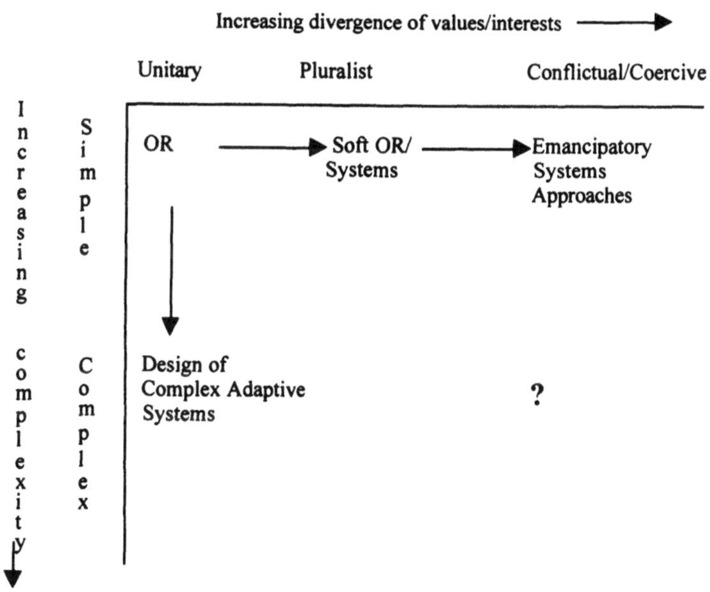

Figure 1. The Development of Applied Systems Thinking

Let's take the discussion on to applications within the social and management sciences, looking at developments in applied systems thinking over the last 40 or 50 years. I like to do this using two dimensions (see Figure 1). Applied systems thinking began with operations research, systems analysis, systems engineering, and management cybernetics.

The emphasis was upon establishing an agreed goal and developing systems of control that would optimise the system towards achieving that goal. This could be achieved by the mathematical modelling of the system of concern, which was therefore treated as being simple enough so that it could be mathematically modelled and optimised in pursuit of its goal. These traditions of systems thinking have been very productive of insights to people who have wished to improve social systems.

We would not have got very far, however, if systems thinking had remained restricted by that particular pair of assumptions, in the top left hand corner of Figure 1 and labelled unitary and simple. It has been an achievement of systems thinking that we have been able to break with these assumptions along two dimensions. Along the horizontal dimension we can now cope with aspects of pluralism such as differences of values, beliefs and philosophies and are able to recognise, at least, the possibility of conflict, power and coercion. Along the vertical dimension, traditions of systems thinking have grown up which have enabled us to deal with greater complexity and turbulence. Progress along both of the dimensions has required philosophical and sociological shifts; epistemological breaks away from the original assumptions of 'hard systems thinking'. Embracing different philosophies and different theories of knowledge in order to make these breaks has allowed us to start to tackle, in applied systems thinking, greater complexity and greater pluralism and conflict.

Specifically along the vertical dimension, a shift from positivism (the attempt to look at systems and trace the relationships between surface variables to try and understand exactly how they behave) has occurred, to what I regard as a kind of structuralism (which is a digging beneath the surface to look at 'deeper' features of systems, the mechanisms at play within systems which lead to surface behaviour). If you can identify these you can start to design those features into systems so that they are capable of complex and adaptive behaviour in relation to their own parts and the environments in which they exist. That is an epistemological break from positivism to structuralism. The kind of systems approaches which achieve this break are system dynamics, looking at the relationships between feedback variables below the surface, socio-technical systems thinking, organisational cybernetics, living systems theory. In terms of metaphor a shift away from looking at organisations as machines or as simple organisms, towards looking at them as 'organisms with brains', has allowed that shift to occur.

To make progress along the horizontal dimension it was necessary to give up the idea that you can make one model of the system of concern. It is necessary to accept and deal with a variety of viewpoints, beliefs, and philosophies about the nature of the system and what it is seeking to achieve. As a result you have to seek to bring about a consensus or accommodation among the variety of stakeholders concerned. Only then can you decide what it is possible to achieve and ensure that there is sufficient agreement between the various stakeholders so they can take effective action. It is the 'soft systems' tradition of work that has had the greatest success in this area. It concentrates upon systems in the mind, rather than systems out there in the world. Emphasis shifts from modelling systems in the world to systemically learning about the nature of what kind of change is possible. So you engineer not systems but learning systems whereby you can arrive at some kind of accommodation on the basis of which action can be taken. This tradition of work is expressed in Warfield's 'interactive management', Ackoff's 'interactive planning',

Churchman's work, Mason and Mitroff's work, and the writings of Checkland in the UK on soft systems methodology.

During the 1980s, applied systems thinking was able to break away from hard systems thinking and to spread its wings. This has enabled us, as applied systems thinkers dealing with social and management issues, to address a much wider range of problem situations with some degree of success, though we still need to be modest in the claims that we make.

PRESENT SYSTEMS THINKING

I used the grid of Figure 1 to write a book on systems methodology applied in the management sciences in 1991 (Jackson, 1991), and in revisiting that volume for a new and extended version of the book, which was published in October 2000 (Jackson, 2000), I have had to take into account more recent developments within the systems tradition of work that could be applied to social systems. Although still in the past a little bit, these are the things that have come to the fore, in my mind, in the last 10 years since the original version of my book. Some of them predate 1990 but my argument is that they came to the fore in the 1990s. We need to note these more recent developments and see how they fit into the picture.

The theory of autopoiesis, introduced in biology, has been transferred to the social domain with some encouragement from Maturana. There is interest in applying it to management these days, with books written about the relationship between autopoiesis, organisational design and organisational epistemology, although with few concrete examples. Its greatest advocate in the social sciences has been the German sociologist, Luhmann.

Senge's (1990) book, 'The Fifth Discipline' became a best seller. We owe a great debt to Senge for popularising systems ideas to managers and his book has obviously been significant over the last 10 years. It is a kind of combination of system dynamics with various soft systems approaches. It is pretty revolutionary in its way to build on system dynamics, to try to grasp the complexity of systems and the way they function according to the feedback variables and the relationships between them at the level below the surface, and at the same time to talk about changing our ways of thinking, getting away from group-think in order to develop 'learning organisations'. A pretty revolutionary book but unfortunately, in my view (Jackson, 2000), with some basic contradictions at its heart.

Chaos and complexity theory I have mentioned as having come to the fore as a candidate for a general system theory. Many books now are trying to popularise chaos and complexity theory in the management and social sciences. There are some interesting ideas coming forward. The question is are we getting scientifically testable isomorphisms? I do not think we are, there is no evidence yet that that is the case. Are we getting useful analogies? Not yet, there is still some more work to be done on that. Are there some useful metaphors? Well yes, there are, chaos and complexity theory does provide some metaphors which can be taken over to management and organisation theory and used in those disciplines.

Soft OR, soft system dynamics, soft cybernetics are all seeking to break away from the original positivist or structuralist routes by taking OR, system dynamics and cybernetics, and seeking to use appropriate models, methods and techniques within a softer framework to help articulate discussions and languaging about the nature of social organisation. Though it pre-dates this thinking, I could mention John Friend's work in soft OR (Friend and Jessop, 1969). In soft system dynamics, there are attempts to use models as the basis for discussion and debate rather than as attempts to model the world. In soft cybernetics, the break with the first cybernetics and the establishment of a second cybernetics, concerned with the cybernetics of the observing system, marks a shift from functionalist to interpretive foundations for cybernetics.

Some of the older strands of work have also moved on. Soft systems methodology now operates according to 2 modes of thought, Mode 1 and Mode 2. Mode 2 is much more connected to the world of the manager. The manager carries out his/her ordinary, everyday tasks and uses soft systems methodology to reflect upon and to enrich her thinking about how she is doing her ordinary managerial work.

Critical systems thinking, wherein I locate my own work, has over the last 10 years helped establish another strand of systems thinking, which I call emancipatory systems thinking. This aims is to deal with all those issues which are ignored in functionalist and interpretive approaches; that is, by the complex adaptive system and the soft systems strands. Issues of inequality between classes, sexes, those with and without disabilities, those with different sexualities, etc., are highlighted in emancipatory systems thinking. Empowerment and emancipation is sought for those groups discriminated against by the way our current systems operate.

Critical systems thinking goes far beyond just this however. It also talks about how we can use systems theories in combination, and methodologies and methods in combination. It seeks to draw upon functionalist systems thinking, the softer systems approaches and emancipatory systems approaches and to put those sets of approaches to work in a coherent attempt to address the problems of social systems from a management (broadly defined) perspective. Multi-methodology and multi-method practice; pluralism, in the use of systems approaches and systems ideas, in the social and management sciences is what is promoted by critical systems thinking.

I cannot ignore the impact of postmodernism. It is on the front of a Toronto paper today. An article talks about Clint Eastwood as a postmodernist. This tends to suggest how pervasive the topic of postmodernism is. Clint Eastwood is related to Derrida, and seen as deconstructing the macho notion of the cowboy. Postmodernism is having an impact on systems ideas. You can carry out systems studies in the spirit of postmodernism; Foucault, after all, was a professor of 'systems of thought'. There are also deconstructive methods which can be employed as part of the systems toolkit in any kind of methodology and can contribute to critical systems thinking in that way. There is a hell of a lot to build on.

The functionalist, interpretive, emancipatory and postmodern paradigms dominate social science today. Still significant is the huge functionalist tradition of work based upon positivism and using mechanical and biological analogies. There is an interpretive tradition of work much more concerned with 'understanding' from the perspective of the human subject; getting inside the head of the subject to understand how he attributes meaning to his situation, how humans together create social reality. Social reality is not seen by interpretive thinkers as a dominating, restraining force. Rather social reality is something constructed by people in everyday interaction. This is the position adopted by Weber and Schutz, as opposed to the functionalism of Pareto and Durkheim. The emancipatory tradition, from Marx, to critical theory, to Habermas and beyond remains influential. And the postmodern tradition of Foucault, Derrida, and Lyotard, is having its day. These are the traditions that dominate the social sciences.

FUTURE SYSTEMS THINKING

I do not think that systems thinking can escape from these paradigms. It is not my finding, from reading the systems literature, that we have escaped from these basic divisions in the social sciences. Instead when I read attempts by systems theorists to study social systems, or to sort out how to intervene in social systems, they are always set within one of these four paradigms. Predominantly the functionalist tradition holds sway. It is reflected in general system theory, organisational cybernetics, living systems theory, OR, systems engineering, system dynamics. The great bulk of work in systems that seeks to address social problems is stuck within the functionalist paradigm. There is a strong interpretive strand, as in the soft systems approaches of Warfield, Ackoff, Checkland and Churchman. The emancipatory tradition is still in its infancy, but consists of the work of Ulrich and ideas derived from Freire and Habermas. Postmodern systems thinking derives directly from the postmodern movement in the social sciences.

The systems tradition of work does not escape these basic divisions, but it does add something which is genuinely important. These are the things that I think it adds and therefore these are the things which will enable us to contribute, as systems thinkers, in the future to the social sciences and to addressing management problems.

We take a holistic approach. That must be taken as read; we are systems thinkers and we take a holistic approach. This does not mean we try to include everything in our systems studies. We know that is what we would need to do to be comprehensive, but we also know that is impossible. So what holism means for us, I hope, is that we reflect upon our lack of being able to address the whole system. Where do we draw the boundaries? What are we leaving out? What are we including? What are the implications in terms of the intervention we are undertaking of where we have drawn the boundaries? Holism is an extremely important idea which is relevant whatever tradition of the social sciences you embrace.

Another idea that systems thinking adds concerns the process of acquiring and constructing knowledge. Systems thinkers emphasise the need to build knowledge on the basis of cognitive systems. We think in systems terms; we want to be coherent in our knowledge. The systems tradition of work can enable the various types of social science to be more rigorous in the concepts that they employ and in knowledge construction. We have already provided a number of concepts to the social science traditions. If we can continue to do that, and make their work more rigorous and more comprehensive, then we are making a significant contribution.

Transdisciplinarity is another area in which systems thinkers lead. Systems is a transdiscipline. We look for useful analogies and metaphors that we can take across the disciplines. I think that chaos and complexity theory, for example, although interpreted largely when applied to management in a functionalist way, has a lot to offer as a new source of metaphors for the social and management sciences. So has autopoiesis. There is still much to be gained from a fluency which enables the switching of analogies and metaphors across the disciplines, and we as systems thinkers are best equipped to do that.

Fundamentally, however, it is in relevance to practice where we can make the most difference. If you ask a functionalist social scientist or for that matter an interpretive, emancipatory or postmodern thinker: "You, think of society that way, now what are you going to do about it in order to intervene in it on the basis of that sets of assumptions?", they largely cannot tell you. Systems thinking has gone further in terms of its ability to translate theoretical ideas down to practice. Of primary significance here are the methodologies we have developed. These are the vehicles whereby we can translate theoretical notions, enriched by systems thinking, from the functionalist, interpretive, emancipatory and postmodern forms of social science down into practice. We have the methodologies to transfer the insights of functionalist approaches down into practice and interpretive emancipatory and postmodern approaches as well. In forging this link between theory and practice we can contribute the most as systems thinkers. Put alongside our holism, our commitment to thinking in terms of cognitive systems and our transdisciplinarity, this is an impressive list of contributions that systems thinking can make to the social and management sciences.

Let me try to draw some of this together using an idea of how any research needs to be conducted, derived from Checkland and Holwell (1998). You need frameworks of ideas, which I think the social sciences can provide, and it will provide these in terms of the broad divisions that we have set out - functionalism, interpretivism, emancipation and postmodernism. If you want to apply these ideas, you need methodology and we have in systems thinking a wide range of methodologies which reflect these broader frameworks of ideas. We also have many methods, models and techniques within systems thinking which we can use as parts of these methodologies. Using these methodologies, based upon the frameworks of ideas, we seek to address particular areas of concern in the social and management domain. As applied systems thinkers we put our emphasis necessarily on areas of concern. We are concerned with organisational design; with how information systems can contribute to organisations; with improving life in organisations; with getting rid of unnecessary inequalities; with getting rid of poverty; with problems of pollution. We are dealing with issues of globalisation and underdevelopment. To applied systems thinkers, the area of concern is paramount. However, it is impossible properly to address, and learn about how to address these areas of concern, without also understanding the methodologies we are using and the frameworks of ideas that we are bringing forward, as candidates, to try to explain some of the things that are going on in those areas of concern.

So it is the establishment of systems research traditions which correspond to this research model, presented by Checkland and Holwell, that is important. There are some very significant and important research traditions within systems thinking which do approximate to this model: The OR research tradition, the organisational cybernetic research tradition, the soft systems tradition, the living systems theory tradition, the critical systems theory tradition of work. These are extremely significant research traditions of work which, when they are at their best, operate according to the model. They address areas of concern, they know what methodologies they are bringing to bear, they know what frameworks of ideas are being encapsulated in those methodologies. As a result you get learning about the frameworks of ideas, about the particular social theories you are employing, about the methodology and the best way to apply those ideas, and about the area of concern that you are dealing with – whether it be organisational effectiveness, poverty, pollution, or inequality. You can create a successful learning cycle.

The claims we are making for the role of systems thinking in the management and social sciences may be more modest than those outlined, by Len Troncale, for systems thinking in the physical and biological sciences. But the problems that we are dealing with are just as important and we should be doing everything we can as systems thinkers to contribute to their resolution.

Acknowledgement

I would like to thank Doreen Gibbs for transcribing this from the original talk. Without her considerable efforts this paper would not have emerged.

REFERENCES

Capra, F., 1996, *The Web of Life: A New Synthesis of Mind and Matter*, Flamingo, London.

Checkland, P.B., 1981, *Systems Thinking, Systems Practice*, Wiley, Chichester.

Checkland, P. B. and Holwell, S., 1998, *Information, Systems and Information Systems*, Wiley, Chichester.

Friend, J. K. and Jessop, W. N., 1969, *Local Government and Strategic Choice: an Operational Research Approach to the Processes of Public Planning,* Tavistock, London.

Jackson, M. C., 1991, *Systems Methodology for the Management Sciences*, Plenum, New York.

Jackson, M. C., 2000, *Systems Approaches to Management*, Kluwer/Plenum, New York.

Senge, P. M., 1990, *The Fifth Discipline: the Art and Practice of the Learning Organization*, Random House, London.

Stacey, R. D., 1996, *Complexity and Creativity in Organizations*, Berret-Kohler, San Francisco.

Troncale, L. R., 2000, The Future of the Natural Systems Sciences, in *Understanding Complexity*, Ragsdell, G. and Wilby J. (eds), Kluwer/Plenum, London.

Wheatley, M. J., 1992, *Leadership and the New Science: Learning About Organization from an Orderly Universe*, Berrett-Koehler, San Francisco.

SYSTEMS THINKING FOR THE 21ST CENTURY

Gerald Midgley

Centre for Systems Studies
Business School
University of Hull
Hull, HU6 7RX, UK.

INTRODUCTION

As we enter the 21st Century, we are witnessing a number of phenomena that provide fertile ground for the growth of systems thinking: a greater recognition of interconnectedness than ever before; scepticism amongst Western populations about supposedly value-neutral science; and a political interest in addressing social exclusion. This paper outlines a new approach to systemic intervention that gives people a useful language to both describe these kinds of issues, and deal with them in a participative and creative manner.

This paper focuses on some new theoretical work that I have been engaged in over the past couple of years, which will be published in a book (Midgley, 2000) more or less simultaneously with the volume you are now reading. Obviously, reducing from a full-size book will involve some radical simplifications, so please forgive me if I surface more questions than answers.

Essentially, I am optimistic about the future of systems thinking. I believe that the conditions we find around us in the West as we enter the 21st Century are making people take increasing notice of systemic intervention. I am therefore going to start this paper by picking out just three issues that make systemic intervention particularly relevant at the present time. Then I am going to outline the systems perspective that I have been developing, that I believe is particularly responsive to the conditions we find ourselves in. This is based on the tradition of Critical Systems Thinking (see Flood and Jackson, 1991, and Flood and Romm, 1996, for reviews).

I will start my discussion of this systems perspective with some *process philosophy*—which is essentially about the process by which any knowledge or understanding comes into being. I'm going to argue that all knowledge is dependent on processes of making boundary judgements, and therefore the 'boundary' concept has to be

Understanding Complexity, Edited by Ragsdell and Wilby,
Kluwer Academic/Plenum Publishers, New York, 2001

placed at the very heart of systems thinking. Next, I will introduce *systemic intervention.* Intervention is purposeful action by an agent to create change, and the way I define *systemic* intervention will follow from the process philosophy I will already have outlined.

This will give me a foundation to tell you a little about the *methodology* of systemic intervention that I have been developing, starting with the theory of *boundary critique*. Boundary critique involves being critical about value and boundary judgements during intervention—thinking about the inclusion, exclusion and marginalisation of stakeholders and issues. I also welcome theoretical and methodological pluralism, which grants some flexibility and responsiveness in interventions.

Finally I'll end by looking back at the issues I introduced at the beginning to show why I believe that this sort of systems perspective is able to address some of the key questions of our time.

KEY ISSUES AT THE TURN OF THE CENTURY

Let me start by looking at what I see as some of the most important conditions we are facing as we enter the 21st Century.

Interconnectedness

The first is *interconnectedness*. In some ways I really don't have to say much about interconnectedness, as systems thinkers have been arguing for many years that the more complex our problems become, the more important it is to take a systems approach. This interconnectedness (for example, between ecological and social issues) has been recognised for a long time in relation to global issues. Nevertheless, the difference I'm now seeing in my own work is that everybody is talking about interconnectedness at a *local* level too. Organisations have to work in partnership more than ever before, and managing this kind of interconnectedness has really become a big issue.

Scepticism about 'Value-Neutral' Science

Another big issue is *scepticism about supposedly value-neutral science*. There seems to be a public awareness now that scientists don't have all the answers, and maybe they're not even asking the right questions. Some scientists are working on high-profile research that has created real concerns amongst the public. An example is the use of genetically modified organisms in food production. It seems that many ordinary citizens simply don't believe the pronouncements of safety that they're hearing from the so-called 'experts'. The arguments of campaigners that genetic modifications could have unknown systemic effects are really carrying some weight.

It seems to me that there is evidence that this is not just a fashion for science-bashing. In the UK, there has been a change in the attitude of some parts of the scientific establishment too. In 1999, the Economics and Social Research Council (ESRC), which is one of the most influential state-funded scientific committees, published a briefing document that virtually declared an end to the idea that science is value free. In fact, they called on the scientific community to radically overhaul the way science is conducted, and put public participation at the heart of the development of research. It seems to me that the idea of the ESRC saying something like this would have been absolutely unthinkable just 10 years ago.

Social Exclusion

Now, the final condition I want to mention is the political focus on *social exclusion*. This is a big issue in the UK, and I believe it is also being talked about in other Western countries. Social exclusion basically means the exclusion of groups or classes of people from participation in the mainstream of society. This can be because of poverty, geographical location, race, gender, disability—any number of reasons. Of course, there is some scepticism about how much of the talk of social exclusion is just rhetoric, and how much is seeking real change. But in the UK at least, there is certainly a great deal of public sector money going into identifying and dealing with social exclusion, and there does seem to be a wide-spread consensus that it's a genuine priority.

I believe that all three of these conditions (a wider recognition of interconnectedness, scepticism about 'value-neutral' science, and the focus on social exclusion) combine to create an unprecedented opportunity for the systems community. Below, I will describe the kind of approach to systemic intervention that I believe can make a difference given the situation in which we find ourselves. I will start with philosophy, and then go on to discuss methodology. Examples of practice can be found in Midgley (2000) and elsewhere.

PROCESS PHILOSOPHY

One of the big philosophical issues for systems thinkers has always been our opposition to both reductionism and subject/object dualism. Here I will concentrate on dualism, as I believe it is a more fundamental problem (see Midgley, 2000, for my reasoning).

A conventional approach to epistemology (the study of the nature of knowledge) asks, 'if there is such a thing as knowledge, what gives rise to it?' There are obviously many possibilities: an individual consciousness; an embodied mind; an individual linked into a linguistic community; an inter-subjective construction of a social group, etc.—there are as many possibilities as there are theories of epistemology.

Anything that can be seen as giving rise to knowledge can be called a 'knowledge generating system'. The trouble is, if the task of epistemology is to identify a general theory that can specify the nature of knowledge generating systems, then it is very easy to slip into dualism. Somehow knowledge generating systems come to be seen as fundamentally distinct from the knowledge of the world that they give rise to.

However, there is an alternative approach to epistemology. I am going to start with Churchman's (1970) fundamental insight that all knowledge is bounded. In other words, while in reality everything may be interconnected, what we actually know about any situation has limits, and it is these limits that we call boundaries. So all knowledge is dependent on boundary judgements, whether these boundary judgements are implicit or explicit. If we recognise this, then suddenly both knowledge generating systems and the world itself come to be defined in exactly the same manner—through the process of making boundary judgements. Knowledge about knowledge generating systems and the rest of the world has the same character: there is no dualism between the subject (knowledge generating system) and object (world).

Of course there *is* still a *kind* of dualism in here, but it's not the pernicious subject/object dualism that underpins reductionism. Rather, it's a dualism between the *process* of making boundary judgements and the *content* of any analysis, whether it's an analysis of what's in the world or an analysis of knowledge generating systems giving rise to knowledge about the world. What this actually means is that it is perfectly possible to accept any number of theories about either knowledge generating systems or the wider world, depending on our purposes. Each theory will be based upon a different boundary judgement.

Of necessity, there is a great deal of detail missing from this discussion of process philosophy. For instance, it is important to acknowledge that boundary judgements cannot exist independently from knowledge generating systems. I therefore recommend consulting Midgley (2000) for more information.

SYSTEMIC INTERVENTION

Having discussed process philosophy I can now consider the concept of *intervention*. I define intervention as 'purposeful action by an agent to create change'.

Given the process philosophy already described, there are many different ways of describing an agent, depending on the boundary judgements that are assumed. An agent may be seen as a lone decision-maker, an organisation, a group of people, a person using a particular form of language, a community within a particular kind of ecosystem, etc. This takes us back to the idea that it is quite legitimate to choose between different conceptions—to choose between different boundaries of analysis. But however many ways there are of conceiving agents for particular purposes, we can still talk about intervention in a *general* sense as purposeful action by an agent to create change.

So what is *systemic* intervention? I define it as purposeful action by an agent to create change *in relation to reflection upon boundaries*. The link with the process philosophy I defined earlier should be clear: if knowledge is dependent on boundary judgements, then reflection on these gives rise to new possibilities for knowledge to inform action.

Observation as Intervention

Now, many authors have championed the cause of intervention, but where I depart from some of these (e.g., Seidman, 1988), is in the opposition they create between intervention and the kind of observation that is undertaken in science. My own view is that there is no such thing as absolutely objective observation, so when scientists set up observations they are actually conducting interventions! This might be a controversial assertion, so let me explain using five short points which indicate the dependence of observations on the positioning of the observer:

First, the systems principle of interconnectedness suggests that all observers are either directly or indirectly linked with what they observe. In other words, there can be no such thing as *truly* independent observation, even though scientists use methods to minimise their influence on what they observe.

Second, observers make *decisions* about what to observe. So scientists can choose what to study, depending on their values.

Third, observers select *concepts* to guide observation. So the same scientific observer may observe in a very different manner depending whether she chooses, for example, reductionist scientific concepts or systems concepts.

Fourth, the *interpretation* of sense data is integral to observation. I simply cannot agree with Popper (1959), for instance, that observation is identical for all people simply because it is a biological process. The empirical evidence tells us something quite different (Bagby, 1957).

Fifth, observations are related to meanings that are generated by communities of practice. Scientists don't conduct their observations in a vacuum: they conduct them in order to make a meaningful contribution to a scientific debate. Therefore, their observations will be constructed in ways that will support them in making a difference in the debates in which they are engaged.

So, we see that scientific observation is undertaken *purposefully*, by an *agent*, to *create change*—even if it's just a change in a scientific debate. Notice that these words precisely reflect my definition of intervention (purposeful action by an agent to create change), making it clear that scientific observation is indeed interventionary.

BOUNDARY CRITIQUE

At this point in the paper, we've had a look at what systemic intervention means (purposeful action by an agent to create change in relation to reflection upon boundaries), and I've argued that scientific observation should be regarded as part of intervention practice. What I want to do now is begin to deepen our understanding of what it means to reflect on boundaries in the context of intervention. I'm going to propose a theory of *boundary critique*.

I've coined the term 'boundary critique' to consolidate the work of a number of authors who have written about boundary judgements since the 1960s, starting with Churchman.

Churchman's (1970) basic insight was that boundary judgements and value judgements are intimately linked. Values direct the drawing of the boundaries that determine who and what is going to be included in an intervention. But taken-for-granted boundaries also constrain the *values* that can possibly emerge. So there's a tight relationship between boundaries and values, and reflection on different possibilities for making boundary and value judgements becomes imperative if an intervention is to be regarded as systemic. This, to me, is the core concern of the methodology of systemic intervention.

Then other authors came along after Churchman and put a slightly different slant on this work. For example, Ulrich (1983) argued that boundary critique should be a *rational* process. To Ulrich rationality is essentially about *dialogue*, so a boundary judgement is only rational if it has been agreed in dialogue with all those involved in and affected by an intervention.

What I've done myself is take some of the insights from Churchman and Ulrich and ask a slightly different question: 'what happens when there are conflicts between different stakeholders making different value and boundary judgements?' My answer is expressed in Figure 1.

In the Figure we see two boundaries, and each of them has a set of values (or ethics) associated with them. We can call the narrower boundary the *primary boundary*, and the wider one the *secondary boundary*. Between the two boundaries is a *marginal* area. Within the marginal area are people or issues that are of concern to those operating with the wider boundary, but which are excluded from the concerns of those using the narrow boundary. The two ethics come into conflict, and whatever is in the margins becomes the focus of that conflict.

Now, conflict can sometimes be productive, or when it is judged to be unproductive it can be resolved, but in many social situations this kind of resolution doesn't happen. What happens instead is that the conflict is simply stabilised. The stabilisation happens by the imposition of either a *sacred* or *profane* status on the marginal people or issues.

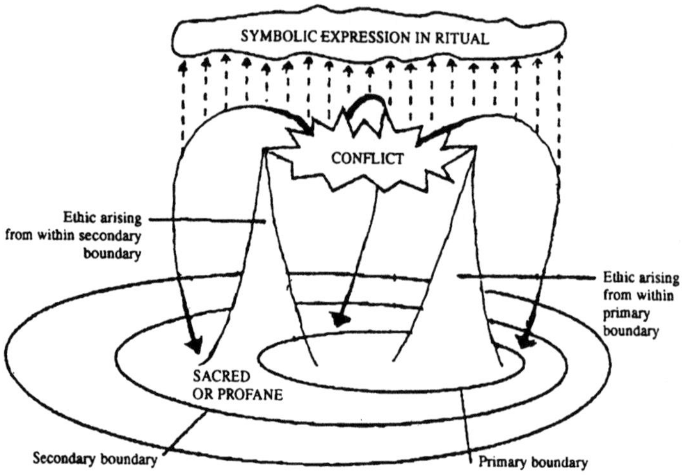

Figure 1. Marginalisation

The terms 'sacred' and 'profane' are not meant in a purely religious sense, but indicate the valued or devalued status of marginalised elements. In fact, there is rarely a consensus about whether marginalised issues or people are sacred or profane, but by institutionalising value judgements in social rituals, the conflict can be stabilised with one set of values dominating. So, if the *profane* status of marginal elements is institutionalised, then the primary boundary is reinforced because people can quite justifiably ignore or derogate whatever is in the margins. But if the *sacredness* of marginalised people or issues is institutionalised, then this challenges the narrow boundary judgement and reinforces the wider secondary one.

As I see it, these kinds of processes operate at every level in society, from small groups to international relations. Many different stakeholders and issues can be marginalised for all sorts of different reasons, and when they are made profane the effects can be quite devastating. Some forms of marginalisation are relatively easy to overcome because they have their roots in very localised histories of conflict, but some stem from conflicts that are really endemic across society, and those are the ones that are most difficult to change.

Some examples of marginalisation (relating to unemployment, customer relations and ecological sustainability) can be found in Midgley (1992, 1994). These may help to further clarify the theory. However, the purpose of presenting this theory is not merely to make an interesting sociological point: I argue that it can usefully inform *intervention*. For details, see Midgley, Munlo and Brown (1998) and Midgley (2000) who give some practical examples of how boundary critique can be operationalised during planning and problem solving.

THEORETICAL AND METHODOLOGICAL PLURALISM

So, I've introduced the idea of boundary critique, which I believe should be central to systemic intervention. Now I want to say a little about the kind of theoretical and methodological pluralism that logically follows on from what I have said about boundaries. Let us start with *theoretical* pluralism.

Different theories assume different boundaries for analysis. If it's legitimate to decide on boundaries out of a wide range of possibilities, it must also be legitimate to draw upon a wide variety of theories.

Now, not only can the systems perspective I am talking about allow *theoretical* pluralism, but it allows *methodological* pluralism too. If we look at different methodologies, we find that they make different theoretical assumptions. So if it is acceptable to welcome in a variety of *theoretical* ideas, a similar variety of *methodological* ideas can have validity too.

In fact, I would suggest that there are *two forms* of methodological pluralism, both of which are essential to systemic intervention. The first is learning from other methodologies to inform one's own. That way, each agent has a continually *evolving* understanding of systems methodology. We no longer have to accept a situation where people build a methodology like a castle and then defend it against others who want to breach the castle walls. Rather, if people begin to see methodology as *dynamic and evolving*, they can learn from others on an on-going basis.

The second form of methodological pluralism is about drawing upon, and mixing, methods from other methodologies (as well as designing one's own). If we can learn from other methodologies on an on-going basis *and* adopt a wide range of methods, then we can have a truly flexible and responsive systems practice.

Arguably one of the great strengths of the systems movement has been the wealth of methods that have been developed to serve different purposes over the years. If we can begin to harness this wealth into a form of systems practice that still keeps the idea of reflecting on value and boundary judgements at its core, I think we will have a very great deal to offer in the coming years. See Midgley (2000) for further details.

CONCLUSIONS

I started this paper by talking about three themes which I see as characterising the point we're at as we enter the 21st Century: an increasing awareness of interconnectedness, scepticism about 'value-neutral' science, and a focus on social exclusion. Let me end by showing how the kind of systemic intervention that I have talked about addresses these themes.

First, interconnectedness. Just about every person who has produced a method for systemic intervention talks about how it can help to *raise awareness* of interconnectedness, or *model* it, or *deal with* it in some way. So methodological pluralism, embracing the widest possible set of methods, gives us an incredible variety of tools to work with the interconnectedness that people are experiencing.

Boundary critique is also useful for this. It is simply not possible to be aware of all interconnections: we always make boundary judgements that *cut* connections, so reflection on boundaries and values is particularly important in the context of intervention.

Next there is scepticism about supposedly 'value-neutral' science. I was very careful when I talked about the meaning of systemic intervention to be clear that methods for guiding observation can be part of intervention too. This means that it is possible to place methods for reflecting on values right next to methods for guiding observation. We can therefore work towards a *value-full* science instead of one that pretends to be value-neutral.

I also believe that there is the *potential* for boundary critique and methodological pluralism to be useful in rethinking the conduct of science. I've been cautious here and talked about 'potential' because it's obviously not been tested yet, and this is a future task for systems thinkers.

Finally, let us look at social exclusion. It seems to me that one of the biggest challenges for systems thinkers in the next decade will be to reconceptualise social exclusion in terms of marginalisation. After all, we're not really talking about *total* exclusion, but the systemic marginalisation of certain sections of society so that they bear the brunt of society's problems. I think that the theory of boundary critique helps to generate some *understanding* of social exclusion, and then being pluralistic about methodology (sweeping in a variety of methods) will help us deal with exclusion in a responsive and locally sensitive manner.

The book you are holding is a millennial celebration of systems science and systemic intervention, and I feel honoured to be asked to contribute to such a volume at a time when systems thinking seems to be entering a new renaissance. Stepping into the third millennium is a *symbolic* event as much as anything, but I get the feeling that it's also a step into a new understanding of the power of the systems idea. We obviously cannot know what the future will hold, but I believe that the kind of systems perspective outlined here will make a significant difference in addressing many of the issues that are now coming to the fore as we enter the 21st Century.

REFERENCES

Bagby, J.W., 1957, A cross-cultural study of perceptual predominance in binocular rivalry, *Journal of Abnormal and Social Psychology* 54:331-334.

Churchman, C.W., 1970, Operations research as a profession, *Management Sci.* 17:B37-53.

ESRC Global Environmental Change Programme, 1999, *The Politics of GM Food: Risk, Science and Public Trust, Special Briefing #5*, University of Sussex, Brighton.

Flood, R.L., and Jackson, M.C., (eds.), 1991, *Critical Systems Thinking: Directed Readings*, Wiley, Chichester.

Flood, R.L., and Romm, N.R.A., (eds.), 1996, *Critical Systems Thinking: Current Research and Practice*, Plenum, New York.

Midgley, G., 1992, The sacred and profane in critical systems thinking, *Systems Practice* 5:5-16.

Midgley, G., 1994, Ecology and the poverty of humanism: a critical systems perspective, *Systems Research* 11:67-76.

Midgley, G., 2000, *Systemic Intervention: Philosophy, Methodology and Practice*, Kluwer/Plenum, New York.

Midgley, G., Munlo, I., and Brown, M., 1998, The theory and practice of boundary critique: developing housing services for older people, *Journal of the Operational Research Society* 49:467-478.

Popper, K.R., 1959, *The Logic of Scientific Discovery*, Harper, New York.

Seidman, E., 1988, Back to the future, community psychology: unfolding a theory of social intervention, *American Journal of Community Psychology* 16:3-24.

Ulrich, W., 1983, *Critical Heuristics of Social Planning: A New Approach to Practical Philosophy*, Haupt, Bern.

OR 2000: CHALLENGES AND PRESCRIPTIONS FOR THE NEW MILLENNIUM

Stig C. Holmberg

Department of Informatics
Mid Sweden University
SE-831 25 Östersund, Sweden

INTRODUCTION

Both Operational Research (OR) specifically, and other Applied System Movements in general have a common concern in dealing with complex and tricky real world situations. Hence, it may be seen as a natural step that The Swedish Operational Research Association (SORA) recently became formally associated with the International Society for Systems Science (ISSS). However, the links between the two societies are far from new. Many individual SORA members are also long-lasting members of ISSS. Further, two of SORA's leading scholars, Professor Kjell Samuelson in the seventies and Professor Donald de Raadt in the nineties, have also served as former presidents of ISSS. Hence, as current president of SORA, I hope that the newly committed association will be beneficial for both our parties and that, together, we will be able to pursue a common endeavour toward a healthier state of world affairs even more effectively.

Hence, as an opening for those forthcoming challenges, I will in this paper both analyse the current status within OR and discuss some possible actions for future improvements. I will basically speak for OR[i] in its broadest sense, but with our shared interest in systemic approaches my conclusions may, to some degree, be relevant also for other branches of the systems field.

ASSESSMENT OF CURRENT OR

So, is OR in a healthy state, is its future development and prosperity secure, and does it provide significant contributions to the improvements of human life in a sustainable world? Further, what main weaknesses and severe deficiencies may be found within OR today? And

Understanding Complexity. Edited by Ragsdell and Wilby.
Kluwer Academic/Plenum Publishers, New York, 2001

finally, what new challenges will OR meet in the near future and what may be improved within OR in order to overcome them? A full answer to those highly relevant but also difficult questions will not be possible here, but anyhow, I will discuss them in the rest of this paper.

Present Status

Today, the origins of OR go back more than 60 years in time and those of Applied Systems Science, or Applied Systemics (AS) just a little less. That earlier development within OR is given an excellent description in, for example, the book *Operational Research and Systems* (Keys, 1991). Klir (1991) with his *Facets of Systems Science,* delivers a similarly comprehensive and thorough overview of Systems Science at large. Both authors, although well aware of some pressing problems, are basically positive to the current state and future prospects within both OR and AS in general. In a recent paper Ackoff (1999), on the other hand, is more critical, especially towards the current state within OR. He even seriously questions if a renaissance or resurrection of OR is still possible! Evidence for both positions is easily found, however, let us first look at the development of OR with help of figure 1.

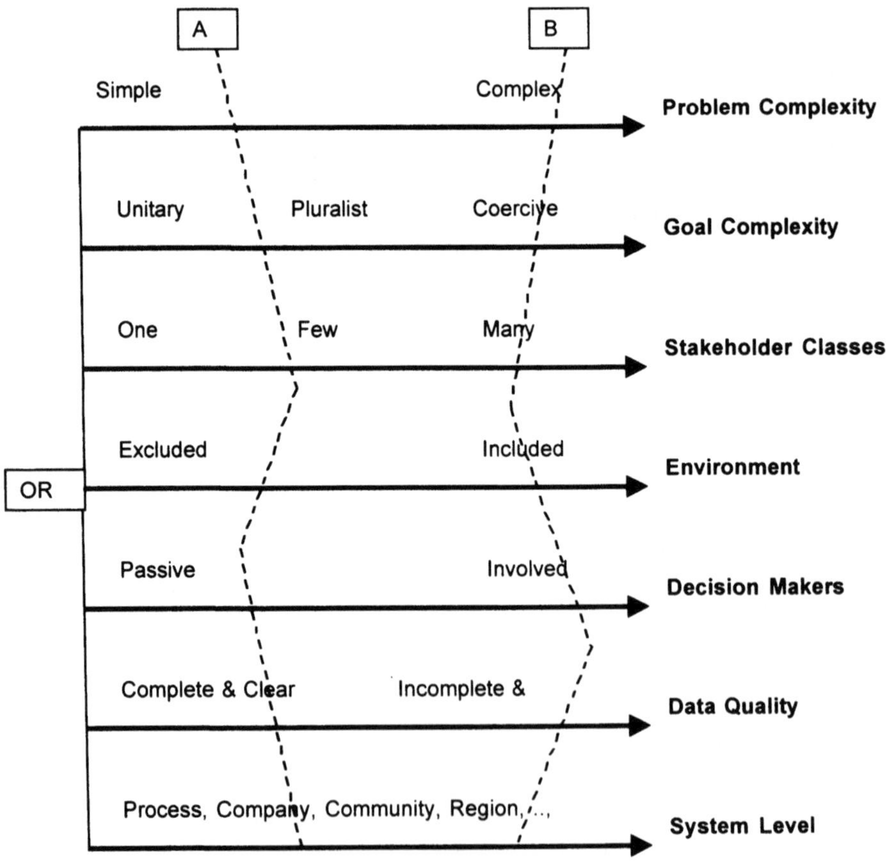

Figure 1. Seven Main Dimensions of OR-tasks and Two Typical Profiles, One Simple and One More Complex.

This figure shows seven main or dominant dimensions of any OR missions. A position to the left indicates a case, which is less complex than a position to the right on respective axis. Hence, the A-profile in figure 1 characterises an OR-task, which is less complex than that of the B-profile. Here, the opinion of Keys (1991) may be understood as if OR generally has moved from left to right over the years. Hence, according to this interpretation, OR gradually has become more systemic and more apt for coping with complex problems. However, according to Ackoff (1999) and to some degree Wilby (1997), on the other hand, OR has moved away from its original intentions of being holistic and systemic. Hence, according to Ackoff (1999), OR has gradually become more and more "*mathematically sophisticated but contextually naive*".

Further, as a movement with active and globally dispersed international and national professional organisations such as IFORS[1] and EURO[2], several high class journals such as ITOR[3] and EJOR[4], and well attended annual international conferences, OR seems to be in a healthy and prosperous state.

However, even here it is possible to find indications of the opposite. For example, in the newly completed National Encyclopaedia of Sweden, i.e. the country's current official "truth", most concepts with a relation to OR and Systemics are defined with rather ridiculing and patronising words. Further, some years ago when Sweden was facing some severe problems, mainly connected to unemployment, renewal of industrial structure and regional balance, the government appointed a group of economists to propose a list of suitable solutions. Everybody found that quite natural and nobody seems to have considered an OR approach and the involvement of OR professionals, not even the OR people themselves.

Hence, even if Ackoff (1999) may be too pessimistic in saying that the future of OR is past, it is evident that its usefulness and versatility is seriously questioned in many circles. The need to improve the status of OR and to increase its usefulness, visibility, and acceptance in society at large is obvious.

Experienced Deficiencies

Undeniably, applications of OR and systems thinking have led to a decent set of success stories. Beside that, a wealth of solid systems knowledge has also been created over the years. Nevertheless, this may not be enough. In looking at the world's dominant survival problems in terms of energy, food, population, environment, conflicts, and so on, at large they seem to be as severe today as 60 years ago. Evidently the efforts of OR and other systems approaches have not changed much. The successes, irrespective of their elegance and sophistication, evidently have concerned only local or minor problems.

Further, in its original intentions OR tried to be holistic, systemic and interdisciplinary, however, lately OR and other branches of Systemics with increasing pace seem to fragment into small and specialised groups. The ordering principle may be a minor set of methods or a bounded set of problems. That has happened in Sweden and probably in many other countries as well. To the degree that specialisation and narrowing of focus is combined with a lack of communication, isolation, and loss of overview, it obviously may become a threat to all basic ideas and ambitions of OR and other system movements.

---

[1] International Federation of Operational Research Societies.
[2] Association of European Operational Research Societies within IFORS.
[3] International Transactions in Operation Research.
[4] European Journal of Operational Research.

Yet another thing, most OR methods seem to have been developed for an abstracted and idealised model world. This can be seen in many basic textbooks, for example that of Taha (1997), where OR is defined and taught in terms of formal methods. Thus, when a student armed with such formal knowledge is confronted with a turbulent, or even hostile, environment and a swiftly changing and messy reality, she or he will find the school book knowledge of limited help. Obviously this deficiency arises from being too focused on formal methods and not enough emphasis on pragmatic and situation specific conditions.

The last deficiency concerns what may be called the OR Trap. This means that OR in its original form is an attempt to solve new and unknown problems with the help of novel, interdisciplinary, and unbounded approaches. However, if such an attempt leads to a new feasible solution, it is tempting to label that as a new OR method and to start teaching it. In so doing, however, you fall into the OR Trap! Any attempt to institutionalise and formalise OR methods is against the original aspirations of OA and hence counterproductive. The fate of real OR is to incessantly seek new solutions to endlessly unfolding new problems!

Challenges for the Future

Obviously all the above deficiencies can be transformed into challenges for future improvements. However, here I will just briefly comment upon three of them.

At first, we may never forget that we still are just in the beginning of the systems era. Compared with other sciences and technologies, OR and other systems approaches are still young and premature. Hence, most of the systems research yet is in front of us, and most of what there is to be known about systems, still is unknown. Hence, OR in the near future has to overcome a huge research challenge.

Next, van Gigch (1996) has observed that there is an obvious contradiction between the working methodologies of OR and the problem domain it is addressing. In other words, apart from a few exceptions, OR has emphasised different versions of rigorous mathematical modelling with its roots in physical sciences, and elderly positivistic and Cartesian thinking. Paradoxically however, most of OR's problem domains are socio-technical systems with inexact and fuzzy properties (Klir, 1996). This opposing nature of OR's problem domain and solution methodology will, inevitably, lead to even more failures in future applications. Hence, the next challenges for the OR community will be to adapt its working methods to its problems and to adjust its epistemological and methodological position accordingly.

A last challenge is the delicate balancing act for systems and operations people to both argue for the strength of the systems approach and at the same time not start overselling the methodology or claiming ability of solving any problem. Being too shy and timid and we will not achieve the visibility and acceptance we deserve. Promising too much, on the other hand, and we will be seen as dilettantes. Jackson (1995) has wittily described this balancing act as a choice between systems thinking and management fads. The challenge, obviously, is to decide on the systems thinking and to always stay serious and rigorous.

POSSIBLE ACTIONS FOR REVITALISING OR

As seen above, the state of OR and related fields is far from unproblematic. Our influence and visibility in society and organisations is in several aspects declining and the severe world problems we are supposed to solve are still rather unaffected by OR activities. However, the situation is in no way hopeless and I can see a lot of potential research and development initiatives in order to improve the current state. Some of those I will shortly discuss here.

Systems Heritage and Systemic Knowledge Support

As already said, we are not able to take care of, and to exploit, the systemic knowledge that has already been produced by our common research community. This sad fact is due to what Herbert Simon (1995) has called the bounded rationality of humans. Hence, in order to master our common systemic knowledge, and other relevant knowledge as well, we would benefit from some sort of support. Today's libraries, databases, and web sites, however, are not able to serve that purpose. They are just excellent data storages. The device I have in mind, on the other hand, will help us see the relations between different research findings and it will make the implications of each clear and evident. Further, it needs to help us identify what is most relevant and important in each situation and it needs to show us what to extract from different data storages. In short, it will expand our short time memory in pushing Miller's fatale number from seven toward 70 or even higher.

Unfortunately, so far I have arrived only with a prototype design of such a device (Holmberg, 2001). However, I find that Stafford Beer's work on InfoSets and Team Syntegrity (Beer, 1994) will form a good platform for fulfilling even this development work.

Here, there is also the problem of introducing new generations into systems thinking and to keep alive the systems knowledge we have already created. Even in this task we could benefit from some form of computer support. During the WC/ISSS'2000 conference Len Troncale (2000) demonstrated the Stealt System in a workshop. Evidently, this type of system could be of great help to the remaining development work and deserves any possible support and encouragement.

System Watch Institute

We have organisations like World Watch Institute, sensing, analysing and reporting what is happening with our entire world. In a similar way I think we need a System Watch Institute (SWI), continuously sensing, evaluating and reporting what is happening within OR and other systems movements. Without such a function, we risk losing overview and orientation. Nobody will have any idea about which initiatives and developments are good or, which are less deserving. Many possible and desirable cross fertilisations and synergy effects risk being missed or overseen.

Such an institute could also point out what we do not know yet, which is rather a lot, and it could define an appropriate and advisable research agenda. Hence, a SWI has an important role for a prosperous development of the systems movement. Russ Ackoff (1976) has already long ago, in a book on a national scientific library system, discussed the possible organisation of such a function. Due to the technological development a realisation today would be even more straightforward and effective than in Ackoff's design.

Development Strategies

In order to promote and strengthen the systems movement I think it would be beneficial to rethink our own development strategies. One thing, for example, is to have common discussions for identifying desirable goals and feasible actions. Alternative idealised futures may be designed and discussed.

In order to achieve full commitment for the results all members ought to be engaged in those discussions and design activities. That in its turn requires special procedures and arrangements. Stafford Beer's Team Syntegrity seems to constitute a well designed protocol even for this purpose (Beer, 1994). Further, Markus Schwaninger (1996) and others have demonstrated that the Team Syntegrity protocol also can be distributed geographically and that it can be extended to a continuous and routine based development activity.

Further, to report on our achievements, we do better by going to non OR and non systems conferences and publish in journals devoted to subjects other than OR and systemics. In that way we will increase the visibility and dispersion of systems work. After all, it is not ourselves we need to convince about the supremacy of a systems approach. Instead we need to meet and influence as many new people as possible.

At last, OR ought to establish better contacts with owners of essential and pressing world problems. Organisations like the above mentioned World Watch Institute and the United Nations could here be relevant examples. In this way OR would be more oriented toward the real world and less toward trivialities, toy-problems, and puzzles. Donald de Raadt (1997) and John van Gigch (1997) give some hints about what such an OR could look like.

This is equally relevant also for OR in industry and organisations. Ackoff (1999) has observed that OR-groups in many business corporations tend to be pushed further and further down in the organisation until eventually they are pushed out. In order to stop this decline, even business OR has to start working more with relevant problems on the strategic level.

Focusing on Design

OR, as with many other system approaches too, have from the beginning been much focused on problem solving or even generic problem solving methodologies. However, several of our most eminent colleagues, for example R. Ackoff, H. Simon and J. Warfield, argue that design is the key issue of most systemic missions. Their main argument being that if we try to identify problems, we also gravely distort reality, and if we try to solve a problem we automatically isolate it from its environment. That means that we become blind and insensitive to all systemic interrelationships.

While problem solving has its focus in the 'here and now', design is more focused into the future. In this respect it is noted that we can never change the history or the current situation but if we arrive to decide on a desirable future state we can also make it come about by working on it from now until that focused time arrive.

There are several good design methodologies and design supports available. Russ Ackoff's (1981) Idealised Design Methodology and John Warfield's (1990) Generic Design with support of the DemoSophia room may be taken as just two good examples. One controversy, however, is to what degree it is possible to design a human activity system in a similar way as you design some sort of machinery or non living artefact. One answer may be that if you have all stakeholders taking part in the design process, for example by applying Stafford Beer's Team Syntegrity (Beer, 1994), it will be possible to build the necessary commitment, in the meaning of Winograd and Flores (1987), for guaranteeing a successful realisation.

Hence, I propose that OR as well as other branches of applied Systemics ought to be conceptualised more as design sciences and less as problem solving methodologies and techniques.

Modalities (de Raadt, 1997) is another concept with strong relations to methodology. Each modality focuses on one essence of reality according to table 1, and in order to seize the reality in its full complexity we need to apply no less than seventeen. De Raadt provides good arguments and, undeniably, the application of several modalities, and the relations between them, give a richer and fuller picture of the reality under focus.

Table 1. Modalities and their respective essence according to de Raadt (1997).

Modality	Essence
Credal	Faith
Ethical	Love
Juridical	Justice
Aesthetic	Beauty
Economic	Viability
Operational	Production
Social	Interaction
Epistemic	Wisdom
Informatory	Symbolism
Historical	Formative power
Logical	Distinction
Psychic	Feeling
Biotic	Vitality
Physical	Energy
Kinetic	Motion
Spatial	Continuous extension
Numeric	Discrete quantity

The Pragmatic Link

A methodology is one thing, its application in a real problem situation is quite another one. If that crucial point is not well understood the consequent disaster will be more or less a certainty.

The link between the two may be called pragmatics. Compared with the methodologies, which by now begin to be rather elaborate and well understood, the pragmatism in methods application is far less developed. Hence, more attention and development efforts devoted to pragmatics or meta methodology would be welcome. My SORA colleague Per Agrell (1991) has for this context developed what we may call a Methodology Matrix. That matrix may be a good help for understanding the different logics, languages, and perspectives needed in a normal systems work, i.e. the application of OR in any turbulent, messy, and swiftly changing environment.

Hence, with insightful pragmatics we can diminish the *thrownness* (Winograd and Flores, 1987) when working with different systemic methods and OR practice may become more robust and effective. In short, pragmatics may drastically increase the success rate.

Methods Reconstructability Analysis

Over time a set of working procedures, problem solving methods and running techniques have been developed both within OR specifically and within Applied Systemics at large. The general development seems to have gone from hard methods over soft methods toward multi or pluralistic ones (Mingers and Gill, 1997). This methodological development may mirror an ambition of OR and systems professionals to engage in increasingly wider and all encompassing questions.

Anyhow, if we assume that there is no single method that is the best one for all problems or situations and that a hard method still may be the best one for certain types of problems and problem environments, there is consequently also a need for some sort of a method selection procedure. Here, evidently there is an interesting parallel with Klir's (1991) reconstructability analysis, i.e. in the same way as we need to test if our models are good enough, we need to be capable of testing whether a method candidate will be suitable for a certain problem situation. To the best of my knowledge, we do not have such a formal test today, but the "Choice Phase" in the "Total Systems Intervention" methodology (Flood, 1995), as well as van Gigch's (1991) work on metamethodology, may form solid launching platforms for the development of such testing tools or procedures. With the possibility of giving a guarantee based on such a formalised certifying process the quality and status of OR work undeniably would increase.

Anyhow, irrespective of the success in this endeavour it is important to remember that new methods and tools will always be developed. Hence, any tendency of isolationism, closeness or imperialism will always impede our working success.

Systemic Modelling

Normally, modelling is a core activity in any problem solving. As discussed by van Gigch (1991), le Moigne (1995) however, modelling is an enormously rich concept. Without going into any details, it is important to observe that modelling is an ongoing process. Hence, the model is not a static representation, which is produced once and for all. On the contrary, it is a relationship between the modeller and the item being modelled. Hence, as our understanding of the situation under study changes and improves, so will the model.

Further, according to both van Gigch (1991) and le Moigne (1995), modelling is always based on, and guided by an epistemology. Hence, a modelling job based on, for example, a constructivist and a positivist epistemology respectively, will yield quite different outcomes. Embarrassingly, many modellers seem to be unaware of this grave fact.

Finally, le Moigne (1995) further argues that there is a fundamental difference between what he calls analytic and systemic modelling respectively. In analytical modelling, analytical and reductionistic working procedures are employed, while systemic insights are guiding systemic modelling. However, even those trying to work systemically paradoxically mostly apply analytical modelling. The reason being, according to le Moigne (1995), that analytical modelling is quite dominant and that the systemic one is hardly known or fully understood on the research level, and much less learned and applied on the practical one. Hence, an effort to fully master even systemic modelling might be very rewarding for the OR field.

Anticipatory Modelling and Computing

Intensified efforts in research and development activities are of paramount importance for the future vitality and viability of OR. In this context however, it is not possible to discuss all potential research with relevance for the OR-field. Let me just shortly comment upon two challenging and promising research areas.

During the latest decennia, at first, several new systems theories and methodologies have been developed. Among those we find for example fuzzy logic, artificial neural networks, evolutionary computing, chaos theory, fractals and quantum information processing. In this context, Klir (1996) says that for example fuzzy logic is not only something useful for system control. On the contrary, it is a completely new way of observing and understanding the world. Hence, it seems a promising research topic to evolve an OR based on those new system theories, i.e. a soft OR based on soft computing. Further, it would be very interesting to investigate how those new theories could support and underpin each other, i.e. what relations and connections may exist between them?

Secondly, Anticipatory Modelling and Computing (AMC) has lately presented itself as a new frontier science (Nadin, 2000), encapsulating several of those new system movements mentioned above. AMC is revolutionary in the sense that future states are taken into account when the next system state is calculated or decided, i.e. an anticipatory system is governed by an anticipatory paradigm instead of the usual reactive one (Rosen, 1985). This falls well in line with Aristotle's Final Causation but sharply breaks with what is normally accepted within natural sciences (Dubois, 1998). Anticipation was earlier seen as strictly bounded to living systems (Rosen, 1985) but Dubois (2000) has lately shown that anticipatory phenomena can be found also in non living, physical systems.

The difference between the two is seen in eqs. 1 and 2. Equation 1 is strictly deterministic and recursive, while the incursive equation 2 even takes future states into account in the computation. This may seem paradoxical but Dubois (1998) has shown how such incursive systems in many cases can be transformed into different sets of recursive algorithms for computation. Hence, the discrete Pearl Verhulst equation may for example be written as an anticipatory incursive equation according to eq. 3. This incursive equation, however, may be transformed into a recursive, non chaotic, equation according to 4.

$$s(t+1) = f[s(t)] \tag{1}$$

$$s(t+1) = f[...s(t-1), s(t), s(t+1),..., p] \tag{2}$$

$$s(t+1) = a\ s(t)[1 - x(t+1)] \tag{3}$$

$$s(t+1) = a\ s(t) / [1 + s(t)] \tag{4}$$

Hence, with an anticipatory model of, for example a business corporation, it would be feasible to test the result of different possible decisions and actions beforehand. Holmberg (2000) demonstrates a first step towards such an AMC based device.

Broadening the Perspectives of OR Practice

Ackoff (1999) and Wilby (1997) criticise OR as being too limited in its problem selection, too narrow minded in its working procedures, and too stereotype in its solutions. Hopefully, this is not the whole truth but, anyhow, it indicates that OR has to develop and improve in various respects. Evidently OR has to move further into higher business firms. Contrary, many problems in the society at large, national as well as global ones, are well suited for an OR approach. Vahl (1996) gives a good example of this new community OR.

However, with this shift in system level, it becomes increasingly difficult to identify isolated and discrete problems. More and more it becomes a web of interrelated and systemic ones. Hence, the design orientation of OR becomes even more relevant and important.

CONCLUSIONS

The Australian quantum physicist Gerard Milburn (1998) concludes that if we accept the findings of quantum physics and the fact that the physical world is a quantum world, then a quantum computer is not only possible, but inevitable. In the same way, as the world is not only quantum but also systemic, systemic thinking and systemic acting will be a certainty with time. Not only within a systems society, but in all parts of human life. Due to this, I am optimistic of the long term future of the OR. The threats OR is facing may be severe but, they are so only in a short time perspective.

Those threats, but also possibilities, concern both internal and external matters. On the internal plane it is mainly a question of method development. They must embrace more dimensions and modalities and become less formal. They ought to be geared more toward design and less toward problem solving. Finally, methods have to be complemented with a pragmatics concerning how to apply them in real life situations. Externally, OR has to be more oriented toward the most essential world problems and it has to become more visible and accepted in the eyes of decision makers and problem owners. In addition, OR must always look for new solutions and new challenges. Starting to institutionalise and exploit past achievements and we will unavoidably fall into the OR Trap!

The remedies I have proposed here may not be the best or most feasible ones but they have demonstrated that there are a lot of possible actions to take. Hence, the crucial point is not the lack of means but if we are willing to raise a common commitment to use them. For the sake of the entire world, which undeniably is in great need of a revitalised and committed OR support, I hope we will do.

Acknowledgements

I want to thank the board members, and the past and present presidents of the International Society for Systems Science for their encouragement and openness in starting up the new liaison and co-operation with the Swedish Operational Research Association. Thanks also to the editors for their kind support and valuable advice during the preparation of this paper.

REFERENCES

Ackoff, R. L., et al, 1976, *Designing a National Scientific and Technological Communication System,* Univ. of Pennsylvania Press, Pennsylvania.

Ackoff, R. L., 1981, *Creating the Corporate Future; Plan or be Planned For,* Wiley, New York.

Ackoff, R. L., 1999, The future of operational research is past, in: Ackoff's Best; His Classic Writings on Management, R. Ackoff, ed., John Wiley & Sons, New York.

Agrell, P. S., 1991, *Systems Theory for Systems Practice,* Report FOA D 10229, Stockholm.

Beer, S, 1994, *Beyond Dispute; The Invention of Team Syntegrity,* Wiley, Chichester.

de Raadt, J. D. R., 1997, A sketch for humane operational research in a technological society, *Systems Practice,* 10:4, pp 421-442.

Dubois, D. M., 1998, Introduction to computing anticipatory systems, *CASYS – Int. J. of Computing Anticipatory Systems,* 2:pp 3-14.

Dubois, D. M., 2000, Review of Incursive, Hyperincursive and Anticipatory Systems – Foundation of Anticipation in Electromagnetism, D. Dubois, ed., AIP Conference Proceedings 517, pp 3-30, American Institute of Physics, Melville, NY.

Flood, R. L., 1995, *Solving Problem Solving,* Wiley, Chichester.

Holmberg S. C., 2000, Designing and prototyping towards anticipatory applications, in: *CASYS'99 – Third International Conference on Computing Anticipatory Systems*, D. Dubois, ed., AIP Conference Proceedings 517, pp 31-41, American Institute of Physics, Melville, NY.

Holmberg, S. C., 2001, An anticipatory searchlight approach, *CASYS – Int. J. of Computing Anticipatory Systems (in print).*

Jackson, M. C., 1995, Beyond the fads: systems thinking for managers, *Systems Research*, 12:1, pp 25-42.

Keys, P., 1991, *Operational Research and Systems*, Plenum Press, New York.

Klir, G., 1991, *Facets of Systems Science*, Plenum Press, New York.

Klir G., 1996, *Fuzzy Sets*, Diderot Publisher, Paris.

Milburn, G., 1998, *The Feynman Processor*, Perseus Books, Cambridge, Ma.

Mingers, J., and Gill, A., (eds), 1997, *Multi Methodology*, Wiley, Chichester.

Le Moigne, J-L., 1995, *La Modélisation des Systèmes Complexes,* Dunod, Paris.

Nadin, M., 2000, Anticipation: a spooky computation. *CASYS – Int. J. of Computing Anticipatory Systems,* 6: pp 3-47.

Rosen, R., 1985, *Anticipatory Systems,* Pergamon Press, Oxford.

Simon, H., 1995, Problem forming, problem finding, and problem solving in design, in: Design and Systems. The International Annual of Practical Philosophy & Methodology. 3: pp 245-258, Collen, A., and Gasparski, W., eds., Transaction Publishers, London.

Schwaninger M., 1996, *To be and not to be, that is the system: A tribute to Stafford Beer.* CD-ROM, Carl-Auer Systeme Verlag, Heidelberg.

Taha, H. A., 1997, *Operations Research; an Introduction,* 6th edition, Prentice-Hall, Upper Saddle River, NJ.

Troncale, L., 2000, The Stealth Systems Science Curriculum, *WC/ISSS'2000 Program and Abstract booklet, International Society for Systems Sciences.*

Vahl, M., 1996, *The Challenge of Community Operational Res*earch. Working Paper no. 9, Lincoln School of Management, University of Lincolnshire & Humberside, Lincoln.

van Gigch, J. P., 1991, *System Design, Modeling, and Metamodeling,* Plenum, New York.

van Gigch, J. P., 1997, The design of an epistemology for the management discipline which resolves dilemmas among ethical and other imperatives, *Systems Practice,* 10:4, pp 381-394.

Warfield, J. N., 1990, *A Science of Generic Design; Managing Complexity Through Systems Design,* Intersystems Publications, Salinas, CA.

Wilby, J., 1997, The observer's role and the process of critical review, *Systems Practice,* 10:4, pp 409-420.

Winograd, T., and Flores, F., 1987, *Understanding Computers and Cognition,* Addison-Wesley, Reading, MA.

ENDNOTE:

Operational Research (OR) is often described as a scientific approach to the solution of problems in the management of complex systems. In a rapidly changing environment an understanding is sought which will facilitate the choice and the implementation of more effective solutions, which, typically, may involve complex interactions among people, materials and money. OR has been used intensively in business, industry, defence, and government. Many new and powerful analytical methods have evolved over the years. OR in practice is a team effort, requiring close co-operation among decision-makers, the skilled OR analyst and the people who will be affected by the management action (from EURO Brochure, June 2000).

Section Eight

ISSS Presidential Addresses

THE SYSTEMS MOVEMENT AND ISSS IN THE NEW MILLENNIUM

G.A. Swanson

Tennessee Technological University,
P.O. Box 5024
Cookeville, TN 38505 USA

Those familiar with the rise of human civilization have little doubt that the emergence of new and enduring sets of relationships and interactions among individual humans is prehistoric. For at least 5,000 years, individuals have introduced novel and innovative ideas and artifacts designed to inter-connect humans in higher orders of life processes. And with every biological generation, different ideas and artifacts were introduced evolving the higher order systems—sometimes to greater effectiveness, sometimes to destruction.

Human inquiry during most of the rise of civilization, in fact, concerned these evolving higher-life processes. Only in the more recent centuries did the main focus of human inquiry change from human systems beyond the organism to the physical and biological systems that support higher-life processes (with the possible exception of the precursors of astronomy and medicine).

The current systems movement, in which the Society for General Systems Research (SGSR) now named the International Society for the Systems Sciences (ISSS) played an important part, arose as a reaction to an overemphasis on analytic procedures of human inquiry which focused on physical and biological systems at the expense of human systems. That which had been largely neglected throughout most of the evolution of human civilization had become the almost exclusive focus.

The peculiar circumstances that gave birth to the SGSR and probably caused its great influence, was the invention of atomic and nuclear weaponry. For the first time in human civilization, a few individuals actually held the physical power to destroy not only the higher-order life processes, but the biological and physical systems that supported them. It was increasingly realized that if advances in the relatively neglected human sciences were not made of similar magnitudes to those made in the natural and biological sciences, the next major evolution of human civilization could be its annihilation.

This reality infused the philosophical goal of an energizing systems movement to reinstate human inquiry by synthesis with the dynamic of service to humanity. I believe it

Understanding Complexity, Edited by Ragsdell and Wilby,
Kluwer Academic/Plenum Publishers, New York, 2001

was that urgency and the simple but epistemologically powerful notion of a system—a set of related and interacting elements—that gave identity to the current systems movement and propelled its ideas into practically every modern academic discipline.

Although its name implied full spectrum scientific investigation and that was intended, the focus of SGSR was clearly theory—general theory. Early on, the General Systems Theory of Ludwig von Bertalanfy was believed to show the greatest promise for the development of an all-encompassing theory—one that could instruct both human and natural systems.

It is particularly the obligation of theoretical science to push successful theories to their failures. Theoretical science makes no major advances until it forces failures. Out of the failures, new advanced theories are proposed and the effort to push the new theories to their failures begins. Theoretical science learns as its theories fail. That is not to say that science advances as scientists fail. The failure of scientists to theorize, test, and apply theories is simply a process of failure, not success. The process of success pushes theory to its failure through testing, development and application.

Over a period of two or three decades, the ISSS and others pushed to its failure the idea that a single grand theory such as General Systems Theory could encompass the human and natural sciences. In doing this, the systems movement reached beyond disciplinal theories that fragment epistemology to meta theories and toward grand theories that, while fostering the investigation of analogy, or isomorphy, of concepts, laws and models in various disciplines, would promote the unity of science and facilitate the transfer of concepts, laws, and models across disciplines.

During the next decade, the ISSS began to self-organize, amidst great fracas, to recognize the importance of general systems theories of varying degrees of abstraction, generality, and purity. Special sessions were organized in annual meetings to study proposed general theories and later to study applications. When such theories were successfully defended and developed, continuing Special Integration Groups (SIGs) were formed to integrate the theories among themselves and with disciplinal theories to apply them to human problems.

The acceptance of general systems-based theories, instead of a grand General Systems Theory, made it increasingly difficult to identify a core of knowledge with which systems science might be identified. The identification of that core, of course, was considered a necessary step in establishing systems science as a viable discipline that could survive amongst the other disciplines of a university-structured knowledge complex. Failing to identify such a specific core of knowledge, we have succeeded to a nondisciplinal process of succession—a process in which no particular theory, though general it may be, has encompassed the universe we have defined for ourselves. That process of succession is the unique characteristic of our society and is our intellectual niche. We failed to find the core of knowledge that we sought because we sought it as specialists among specialists. But in our failure, ISSS emerged as a process of the ever-present, a process of succession, of success. That dynamic will characterize the second wave of the systems movement into the new millennium.

Although we are drawn to ISSS as generalists, our bread and butter jobs are mostly specialties. We are hybrids; and when we move outside our general theories, we often revert to our specialist ways. In our search for a systems science core, we overlooked it because of its simplicity. The core is simply that all things real and imagined may be viewed as systems (sets of related and interacting elements) and, consequently, all things may be known by their relations and interrelations with all other things. But that statement is not enough to define a specialty and so the specialist in us persists in developing a special core of general knowledge.

Our specialist compulsions are not all bad. They serve us well in testing, developing, and applying successful general theories. In fact, without them, we might be content to bask in the infinite leisure of the general and the abstract. Our specialist compulsions draw us to dirty our hands in the complex problems of humanity and to seek solutions through our simplifying general theories. Within ISSS, these compulsions are expressed in the SIGs. Here each general theory is tested, developed, and applied to disciplines, professions, and problems and integrated with other general theories.

In the absence of a specialist-styled core of knowledge, the SIGs act as search engines within the ISSS, providing orderly paths for newcomers to find their way into the diverse but systemically integrated body of knowledge that constitutes systems science. Within SIGs, pre-embryonic systems science cores develop and cross-fertilize with those of other SIGs. I predict that out of these activities, an embryonic science of the general will emerge, drawing together the less general cores of various systems sciences. However, it may well always be a dynamic, evolving core of knowledge.

This system of synthesis is often misunderstood by those members who come and go as their particular interests are stimulated and wane. Many organizations have SIGs, but that acronym stands for special interest groups in most. We all know that special interest groups tend to fragment organizations as well as knowledge. The ISSS is not a conglomerate that grows by fragmenting into special interest groups as members specialize in limited aspects of a discipline. It is the opposite. Through special integration groups, our society synthesizes diverse disciplinal knowledge into cohesive wholes, systemically integrating that knowledge. Curiously enough, as organizations of special interest groups grow by increasing the number of their special interest groups, the ISSS grows as we increase the number of our special integration groups.

These groups form the tentacles that reach into the diverse disciplines of human investigations to build the bridges among them for the unifying of the arts and humanities as well as the sciences. Those bridges are not built from one discipline to another through informal analogy and the forced-fitting of methods and models as often occurs in grand umbrella interdisciplinary organizations. Instead, as 2,000 years ago all roads were said to lead to Rome, the interdisciplinary bridges we build all lead through systems science. And, as with the great transportation hubs of today, you can go anywhere from there. When the bridge is built between systems science and a discipline, the whole world is open to that discipline and that discipline to the whole world.

Please notice that all bridges are built through systems science. Science is our bent. We are a progeny of the American Association for the Advancement of Science (AAAS). We were organized under the History and Philosophy of Science section of that organization. But while science is our perspective, we peculiarly reach beyond science to relate science with the arts and humanities. Each informs the others. This relationship, however, is sometimes misunderstood. Unquestionably, the ISSS is a scientific organization. We seek to advance science. We are not simply an information-processing organization or a political action organization. Science, nevertheless, is concerned with all of existence and our Society is particularly interested in the grand theories that should instruct science and the application of science to human problems. We must interface and interact with the arts and humanities—we must be informed by them and inform them. The problems facing humanity in this new millennium cannot otherwise be solved.

This broad outlook presents us with its own set of problems. Science at the fringe is far less orderly than hard-core science. Science seeks the exact while art seeks the ambiguous. Drawing the two together in the interest of humanity requires tolerance and respect—the artist's tolerance of the scientist for the irritating detail of exactness, the scientist's tolerance of the artist for the exasperating vagueness of ambiguity, and the respect of both that the ideas they strive to communicate across this great chasm are substantive. Only in an environment of tolerance can the grain be separated from the chaff and the grand harvest of success be reaped. We can rather easily agree with the idea of tolerance at this level of abstraction. Getting down to the "nitty-gritty" has too often proven much more difficult.

By what standard do we judge ideas presented? Again, our specialist ways demand the rigor of a refereeing process and, to the extent that cores of knowledge have developed within SIGs and for specific general systems theories, that process may be applied. But new and initially developing ideas do not lend themselves to such processes. There the need for tolerance and respect sometimes demands great faith and our faith is not always rewarded. But we tolerate the idea until it is shown to have or not have substance. Our attempts to introduce specialty developed review processes that apply to all papers submitted to our annual meetings have repeatedly failed. Out of that failure, a system of pushing the review process down to the SIGs and incorporating a wide range of differing decision criteria commensurate with the SIGs' goals is evolving. There the different degrees of scientific rigor of various sciences, as well as the differing logics and ambiguities of the humanities and arts may be assessed. As the editorial staffs of systems journals work with this problem, the process is strengthened. These critical advances will continue.

It is difficult to view educational program failures as advances because they are many times not a process of succession but termination. The systems movement, nevertheless, has contributed significantly to education. Our quest for a core of knowledge may have been driven more by a desire to establish systems science programs in universities and colleges than to conceptualize viable systemic espistemologies and supporting methodologies. Today some herald the demise of several systems education programs in a eulogy of the systems movement. The record should be set straight. The ISSS was not organized as a pedagogical association. It was born in an epistemological quest. We all hope that our ideas will spawn methodologies that, in fact, will produce answers and identify new questions in such volume that a pedagogy to transfer that knowledge will become necessary. But the greater success may be the rapid absorption of our ideas in existing pedagogies.

Pedagogical advances follow naturally epistemological and methodological advances—and, perhaps, premature pedagogical expressions are as natural. We are ever-learning and there is a compulsion to communicate to others that which we learn. It might be expected, then, that systemic pedagogies would spring up before their time to flourish for a while but ultimately fail for lack of demand for the less than matured systems of knowledge they deliver. That process of successive failures is a learning process, and, out of our many pedagogical failures, we have advanced systems pedagogy. When a science of the general emerges, its pedagogical system will follow.

Perhaps the greatest challenge of the new millennium is ISSS' opportunity to provide a professional organization that understands to some degree a pedagogy for the interdiciplinal integration of knowledge. Some new standards for regional and specialty accreditation of higher education programs sound like something we would have written. Integration across disciplines is required and we know that systemic integration of knowledge works. The time is right for a generalist approach to the advancement of systems pedagogy. We should grasp this opportunity even though the integration generally envisioned may fall short of the degree of epistemological support we seek in the full emergence of systems pedagogy.

For more than forty years, the International Society for the Systems Sciences has determinedly, if erratically, pushed its successes to their failures. In 1997, we took the challenge to . . . take it to the limit one more time. Today, ISSS is an important player in the worldwide resurgence of systems thinking. The compelling needs of the new millennium concern higher-order human systems—and require grand systems approaches. ISSS has a contribution to make.

PRESIDENTIAL SPEECH:
THE SYSTEMS SCIENCES IN THE YEAR 3000

Peter A. Corning

Institute for the Study of Complex Systems
119 Bryant Street, Suite 212
Palo Alto, CA 94301 USA

(1950) "Progress is our most important product."
— G.E. Advertisement

(1960) "The future lies ahead."
— Mort Sahl (Comedian)

(1970) "The future is not what it used to be."
— Kenneth Watt (Ecologist)

(1980) "The future will arrive sooner than we expect."
— Lynton Caldwell (Political Scientist)

(1999) "The future may be canceled."
— Eugene Linden (Author)
The Future in Plain Sight

The title of my talk, "The Systems Sciences in the Year 3000," is hyperbole, of course — an exaggeration for effect. Who knows what the world will be like even 100 years from now, much less in 1,000 years? For instance, who can predict the ultimate limit to human population growth, worldwide? Certainly not the demographers, who regularly adjust their forecasts and never seem to agree among themselves.

Or who — even as recently as ten years ago — could have predicted the Internet revolution? Not Bill Gates of Microsoft, who was surprised by it and had to scramble to catch the train. Even today, who can predict the long-term impact of the Internet and

Understanding Complexity, Edited by Ragsdell and Wilby,
Kluwer Academic/Plenum Publishers, New York, 2001

broadband communications, say in 10 or 20 or 50 years? And who would dare to predict what the stock market averages (the Dow Jones or the S&P 500) will be like even one year from now? Once upon a time it was commonplace to make ten-year stock market forecasts, but these have become a joke in recent years. Back in 1990 when the Dow was at 2900, the "consensus forecast" among market analysts for the year 2000 was a lofty 6,000. In fact, it passed through 11,000 earlier this year.

Some of the many wrong-headed technological predictions over the years are legendary. They regularly circulate in e-mail messages these days:

- After evaluating the new "telephone" in 1876, Western Union concluded that "the device is inherently of no value to us."
- Lord Kelvin, the chemist and President of the Royal Society in 1895 declared that "heavier-than-air flying machines are impossible."
- French Marshall Ferdinand Foch (just before World War One) concluded that "airplanes are interesting toys but of no military value."
- Radio pioneer David Sarnoff was told by potential investors in the 1920s that "the wireless music box has no imaginable commercial value."
- Or remember the skeptical Jack Warner, President of Warner Brothers movie studio in 1927, asking "who the hell wants to hear actors talk?"
- Then there was Thomas J. Watson, Chairman of IBM, in 1943: "I think there is a world market for maybe five computers."
- And Bill Gates in 1981: "640K [of memory] ought to be enough for anybody."

Far more significant, though, are the widely varying predictions relating to our global future — the ultimate fate of humankind. At this conference alone, we were offered two very different visions of the future. On the one hand, a multi-media plenary presentation warned us of a potential Armageddon — a war to end all wars in a very different sense from what the American President Woodrow Wilson promised after World War One — if we don't radically change our course. In another conference presentation, however, we were told that human societies are being propelled by an energy-driven pulse — a steep parabola of fossil-fueled growth and inevitable, unavoidable decline — yet a "prosperous" downsizing rather than a great crash is quite feasible.

In sharp contrast, a blithely optimistic vision is being touted these days by the best-selling author Robert Wright (2000) in a new book with a portentous title that was borrowed from game theory, *Nonzero: The Logic of Human Destiny*. Wright assures us that the current era is only the "storm before the calm." A magnificent new structure of global prosperity and world government is being erected before our eyes, he says.

Predicting the future — the ultimate balm against the uncertainties of life — is a very ancient pastime. The Old Testament is full of inspired prophesies. In Western social thought, the philosopher Aristotle was perhaps the first to propose that human societies had an inherent direction or destiny, what he called an "entelechy". Later on, Aristotle's vision was adopted and re-packaged by various Christian Theologians, though with a different end in view.

Much later still, during the European Enlightenment era, "progress" (with a capital "P") became the "consensus forecast" (so to speak). For example, the Marquis de Condorcet, one of the so-called Philosophes, turned the ideal of human progress into a sequence of ten developmental "epochs". Condorcet also had a grand vision — later adopted by Karl Marx — of a future utopia in which material inequalities, class differences and immorality would be eliminated. "The time will come — Condorcet wrote — when the sun will shine only upon a world of free men who recognize no master except their reason, when tyrants and slaves, priests and their stupid or hypocritical tools, will no longer exist except in history or on the stage.

In a similar vein, the 19th century English polymath Herbert Spencer formulated a "Universal Law of Evolution" that encompassed physics, biology, psychology, sociology and ethics. In effect, Spencer deduced society from energy by positing a sort of cosmic progression from energy (which he characterized as an external and universal "force") to matter, life, mind, society and, finally, complex civilizations. Spencer also envisioned a future in which industrial progress would lead to an end of wars and a withering away of the state.

Though the dream of being able to predict the course of the human career retained its seductive appeal in the 20th century, it often seemed as if Cassandra and Pollyanna were at war with each other. On the one hand, we had prophets of doom like the biologist Paul Ehrlich (1972), whose book *The Population Bomb* was a best-seller in the 1970s. Even more disturbing were the gloomy predictions generated by the Club of Rome's *Limits to Growth* models, which produced an impassioned controversy.

On the other hand, "futurists" like Herman Kahn of the Hudson Institute were quite sanguine about having a world population of 15 billion or more, while the well-known anthropologist Raoul Narroll analyzed the historical trend toward ever larger political aggregates and concluded that the probabilities for achieving a world state ranged from the year 2125 (with a 40 percent probability) to 2750 (a 95 percent probability). Anthropologist Robert Carneiro came to a similar conclusion but used a different methodology. He projected forward the historical trend toward a decreasing number of independent political units and declared that, provided nuclear annihilation can be avoided, "a world state cannot be far off." It is a matter of centuries or even decades, not millennia, he declared.

Even the founding father of Sociobiology, distinguished biologist Edward O. Wilson, joined the futurist parade in a book coauthored by Charles Lumsden titled *Genes, Mind and Culture*, (Lumsden and Wilson, 1981) where he advanced what he called an "autocatalytic" theory of cultural evolution. "The ultimate triumph of both human sociobiology and the traditional social sciences would be to correctly explain and predict trends in cultural evolution on the basis of their own axioms....Will the social sciences...be able to explain history more fully and perhaps even predict with moderate accuracy? We believe the answer is yes, at least on a limited scale....The prediction of history is a worthwhile venture."

All the evidence suggests the contrary. Forecasting the future is a very hazardous business, at least if you care at all about being correct. Some of these hazards were addressed at our conference, in one of our World Congress symposium panels. The bottom-line was that forecasting is useful, but we must always be prepared for dramatic surprises — for the unexpected and the unpredictable. This symposium panel reminded me of one of the more memorable statements by the American President, Dwight D. Eisenhower, based on his experience as a wartime leader in World War Two. “Plans are useless,” he said, “but planning is indispensable.”

The problem with forecasting the future is that living systems are not exemplars of ideal types or slaves to linear forces but are messy, historical phenomena. The “caprices” of history are not simply quirks, anomalies or blips; they are not temporary road-blocks that can be got around. They are major causal variables, an integral part of the causal dynamics. In other words, historical processes require a science of history — a science that encompasses and integrates contingent, historical influences. Deterministic, law-driven theories of evolution — or of human history — are destined to fail.

Let me give you one illustration: The computer scientist/psychologist John H. Holland (2000) in his latest book, *Emergence: From Chaos to Order* asks: "How do living systems emerge from the laws of physics and chemistry? Can we explain consciousness as an emergent property of certain kinds of physical systems?" Holland, who is one of the leading figures at the Santa Fe Institute, concedes that emergence can be defined in various ways, and that many aspects remain "enigmatic". He confines his attention only to the sub-set of systems that are "rule-governed" — like chess, he says — systems in which "a small number of rules or laws can generate surprising complexity." Of course, laws don't "generate" anything or "govern" anything. They identify regularities in relationships, and in the patterns of causation in dynamic processes; we characterize these formulations as "laws" because they allow us to generalize and make predictions.

But that aside, Holland's example serves to illustrate the basic problem with a deterministic approach to explaining evolution, and especially complexity. Even in chess (to use Holland's metaphor), you cannot utilize the rules to predict history — that is, the course of any given chess game. Indeed, you cannot even reliably predict the next move in a chess game. Why? Because the "system" involves more than the rules of the game. It also involves the players and their unfolding, moment-by-moment decisions among a very large number of available options at each choice point. The game of chess is inescapably historical, even though it is also constrained and shaped by a set of rules, not to mention the laws of physics. In other words, human knowledge, human creativity, and human folly are integral parts of the historical process, for better or worse.

Nevertheless, there is much to be gained by taking a long view and considering thoughtfully the problem of sustainability for future generations. If we cannot make unequivocal predictions, we can make many useful “if-then” predictions based on specific contingencies. This, I believe, will be the primary challenge for the new millennium, and for the systems sciences. To use Alan Lightman’s image, we are like a speeding train without an engineer. It is time to climb aboard and seize the throttle.

One useful starting point for peering into the future is the concept of synergy. Each of us has an array of basic needs that must, by and large, be satisfied continuously. We cannot, for instance, do for very long without fresh water, or waste elimination, or sleep. Accordingly, each of us — individually and collectively — requires a synergistic “package” of resources and suitable environmental conditions. A society that can reliably provide this package will thrive and possibly grow larger. But if even one of these needs is not satisfied — if any part of the package is deficient — the entire enterprise is likely to be threatened.

Over the course of the past 10,000 years, since the rise of civilization, many new technologies have been developed that enhance our ability to provide for our basic needs (and much more besides). But the Faustian bargain is that we have also become the captives of these technologies. I call it the paradox of dependency. The more valuable is a resource or a technology, the more dependent we become and the greater the cost of losing it. Thus, many of our technologies also amount to "basic needs" that are also vital to our survival. For instance, the very idea of a modern industrial society without electrical power, high-speed transportation and long-range communications is unthinkable.

These bedrock needs frame the nature of the problem we confront, and they allow us to make a great many contingent, "if-then" forecasts. But that is all we can legitimately do. We cannot make unqualified predictions about the future because we do not, and cannot, know in advance all of the variables that will affect its course, including our actions, or inactions.

For example, we can safely predict that there will someday soon be an upper limit to human population growth. As the economist Herbert Stein has observed: "If something can't go on forever, it won't." However, we can only guess what that ceiling will be, much less when and how it will be reached. If by some miracle we were able to institute rigorous world-wide population control measures, our ability to predict global population size would measurably improve.

Nevertheless, we still might not be able to anticipate the negative effects on population growth of such looming threats as global warming, pandemic diseases, monster earthquakes or climate-altering volcano eruptions. (When Mount Tambora in Indonesia erupted in 1816, its enormous dust cloud affected weather conditions and crops as far away as New England for many months afterward. The locals referred to it as "the year without a summer.")

We can also observe many potentially serious future survival threats. If major steps are not taken to address these challenges, then the ultimate consequences are predictable. I will provide just one example: fresh water. As the *New York Times* noted in a recent editorial, "the most precious fluid on earth is not oil, but water." With or without global warming , the world stock of fresh water is being depleted at a rate that will soon threaten our food supply; a major share of the world's agriculture depends upon artificial irrigation. Drinking water and water for sanitation and industrial uses are also threatened. Even now, some 1.3 billion people (20 percent of the global population) do not have safe drinking water, and at least four million people die each year from water-borne diseases. Within the next 50 years, expected population growth is likely to impose a demand for a 50-100 percent increase in fresh water supplies, a staggering challenge. Yet we are currently depleting many of the lakes, rivers and aquifers that serve existing populations. For instance, the great Ogallala Aquifer, a huge underground river in the American southwest (once the size of Lake Huron) that many people in that region depend upon, will go dry in 20-30 years.

To make matters worse, so-called "megadroughts" are a thing of the future, regardless of global warming. We now know that, over the past 10,000 years alone, global climate changes have often led to prolonged droughts that far surpassed anything we have seen in recent centuries. Indeed, the collapse of many ancient civilizations may well have been caused by these severe, decades-long climate disruptions. Yet we are totally unprepared for a recurrence of this likely event.

Consider this: the state of California, with its rich soil, salubrious climate and the longest growing season in the world, produces 90% of the apricots, 87% of the grapes and avocados, 86% of the peaches, 83% of the lemons and strawberries, 80% of the artichokes and lettuce, 73% of the broccoli and 53% of the cauliflower grown in the United States, along with about one-third of the cherries and pears and a significant percentage of the nation's oranges, wheat, rice and other crops. California also currently has a population of about 33 million people that is projected to grow to 49 million by 2025. Unfortunately, California is one of the areas that has been susceptible to severe megadroughts in the past. And, in a food economy that is increasingly global in scope, many other countries besides the U.S. could be hit with shortages and soaring food prices if another such megadrought were to occur in this state.

One solution might be for Californians to move to Canada. (I am being facetious, I hope.) But a more viable option might be to undertake a long-range development program that could "falsify" this prediction (to use the scientific jargon). Solar power is becoming increasingly competitive, and so are windmills. In the 1980s, when there was a burst of experimentation with new energy technologies, the most advanced windmills could produce energy for about 38 cents per kilowatt-hour. Now the cost is down to 3.5-5.0 cents per kilowatt-hour, and wind farms have been cropping up in such unexpected places as the Iowa corn belt. Within the next decade wind farming may become a major industry. Similar improvements have been occurring in solar technology.

So, if we let our imagination range freely, we can foresee a day in the not too distant future when vast arrays of environmentally-friendly solar and wind-powered energy generators could be coupled to improved water desalinization and purification systems. What now seems like a major threat — namely, a growing shortage of fresh water and future megadroughts — could in time evaporate (excuse the pun). But this outcome is not foreordained. The choice is up to us.

The biologist Garrett Hardin, who is famous for his classic article on "The Tragedy of the Commons", (Hardin, 1968) many years ago penned what remains a profoundly important truth about the human condition:

"We cannot *predict* history but we can *make* it; and we can *make* evolution. More: we cannot avoid making evolution. Every reform deliberately instituted in the structure of society changes both history and the selective forces that affect evolution — though evolutionary change may be the farthest thing from our minds as reformers. We are not free to avoid producing evolution: we are only free to close our eyes to what we are doing." (Hardin, 19

As the philosopher John Locke truly observed more than two centuries ago: "Hell is the truth seen too late." I cannot predict that there will be another millennial World Congress of the Systems Sciences 1000 years from now, but I am an optimist. So I would like to invite your descendants to attend the 1044th annual meeting of the International Society for the Systems Sciences, a mere 50 generations from now.

However, the main point of my speech is that we need to begin planning for that event now. Our challenges are immense. But so are the opportunities, if we can seize them before it is too late.

REFERENCES

Club of Rome, 1972, *The Limits to Growth: Report for the Club of Rome's Project on the Predicament of Mankind,* D,H, Meadows et al., eds., Pan, London

Ehrlich, P., 1971, *Population Bomb,* L. Ballantine for the Friends of the Earth

Holland, J.H., 2000, *Emergence: From Chaos to Order,* Oxford University Press, Oxford

Hardin, G., 1968, The tragedy of the commons, *Science,* 162:1243-1248

Lumsden, C.J., and Wilson, E.O., 1981, *Genes, Mind and Culture: The Coevolutionary Process,* Harvard University Press, Cambridge, MA

Wright, R., 2000, *Nonzero: The Logic of Human Destiny,* Little, Brown and Company, UK

INCOMING PRESIDENT'S ADDRESS
CONTINUING THE TRADITIONS OF ISSS:
SYSTEMS SCIENCE IN THE SERVICE OF HUMANITY

Harold Nelson

Systems Design
Antioch
Seattle, WA

INTRODUCTION

The society, founded in 1954 by Ludwig von Bertalanffy, Kenneth Boulding, Ralph Gerard, and Anatol Rapoport as the Society for General Systems Research (SGSR), was created in reaction to the growing disparity between reductionist science and their own scientific understanding of the real world as based on relationships, interrelationships and emergent qualities. Their intention was to integrate different fields of study in the same way that the concept of consilience strives to unify all rational knowledge. Because of this, a role that has been unfairly attributed to ISSS and other systems organizations in the past is one of striving for control and totalization yet this is clearly not the intention of the Society. This misunderstanding however reveals a root concern for global consequences due to localized actions as well as issues of power and authority in contexts of diversity. This tension between global systems and locally autonomous systems will be one of the focuses of the ISSS conference in South Africa next year.

The Society is becoming more successful at demonstrating to colleagues in traditional disciplines and to the public at large that systems thinking provides an important and essential understanding of the world in all of its interrelated complexity. The evolving challenge now is to understand how to utilize the insights gained from systems scholarship in broader social realms. As a consequence, one of the enduring contemporary roles that ISSS has taken on is that of **service** and the one that I want to continue supporting during my tenure as President of ISSS. The role of **service** has been an explicit theme for several past presidents including C. West Churchman, Bela H. Banathy and our most recent president, Peter Corning. Progress in this area is essential for the continued success of the Society in the future.

Understanding Complexity, Edited by Ragsdell and Wilby.
Kluwer Academic/Plenum Publishers, New York, 2001

Acknowledgments and Thanks

I would like to thank Peter Corning for the exceptional job he has done as President in putting together not only the 44th Annual Meeting of the ISSS but the World congress of the Systems Sciences here in Toronto as well. It has been an exceptional gathering of scientists and scholars from diverse systems organizations from around the world. I would also like to thank Helmut Burkhardt and his staff from Ryerson Polytechnic University for their hard work in pulling everything together here on campus in support of the conference attendees and activities. I want to thank Jennifer Wilby and Janet K. Allen for their superb work in publishing the Proceedings of the World Congress and the ISSS 2000 together plus all their invaluable staff work prior to and during the conference.

2001 Conference Announcements

I would like to invite you to attend the 45th international conference to be held at the Asilomar Conference Center, Asilomar, California, July 8th through the 13th, 2001. This will be an opportunity to meet new colleagues from diverse and proud cultures who represent rich historical traditions. The conference will include activities at distributed sites in South Africa, China, Canada, the United Kingdom, Bulgaria, and Peru.

I invite you to consider submitting papers for this conference especially with a focus on the theme of service that I will discuss in greater detail later. The two questions I want to pose for the conference that emerge from the general theme **of Systems Science in the Service of Humanity** are:

1. How can the systemic relationship of *service* serve an emerging understanding of development that is both global and local?

2. How can systems scientists, scholars and practitioners *serve* on behalf of others?

But first I would like to discuss some focus issues that emerged for me during the past few days of the conference. I believe these focus issues ought to become part of an ongoing dialogue within ISSS and invite others to suggest ways for this to happen in a manner that assures action items will emerge in time to be presented at the next conference.

Focus Issues From Conference Sessions

1. Women and Other Under-Represented Populations in ISSS. One of the key issues that emerged during the Congress and ISSS Conference concerned not what was happening but what was not happening. It was clear that there were not enough women making presentations and in organizational positions of leadership. It is important to determine why this has occurred and how ISSS can facilitate more balanced programs in the future in terms of gender participation. There is also a need to develop ways to become more representative of the diversity of populations that exist around the world. This will be especially critical for ISSS as it plans for its next conference. It was also clear that there were not enough students participating actively in the conference. Students are not only important to the future of ISSS but are important to the health and energy of the Society today.

2. Intellectual Skunk Works. Another topic that came up during the Past President's Roundtable discussion, which I had the honor of chairing, was a valuable reminder of the role that ISSS has played historically in providing a context for scholars whose ideas were well ahead of their time. The discussion brought out examples of now-famous thinkers and their ideas that were 20 to 30 years ahead of their time when ISSS was the only professional society willing to give them a forum from which to be heard and published. I believe that it is important for ISSS to remain an intellectual skunk works for new ideas of excellence, especially as we struggle with the issue of defining and confining the field of systems to clearly delineated domains and concepts.

3. Unity in Diversity – i.e. SIG's and the Society. A second issue that arose during the roundtable dealt with the perennial issue of unity in diversity as represented by the SIG's. Although SIG stands for Special Integration Group, rather than Special Interest Group which is common in other professional organizations, there is an ongoing tension between the purpose of ISSS to find common ground among diverse thinkers and the need for individuals to find an intellectual home in domains of interest populated by colleagues with shared intellectual concerns. I believe it is important that this issue be looked at in greater depth in order to find a way to accommodate diversity in unity and unity in diversity.

4. Quality – Rigor and Relevance. A third issue which emerged during the roundtable discussion was about an ongoing concern in ISSS which is the quality of papers and presentations. Quality standards similar to other professional groups such as AAAS (American Association for the Advancement of Science) were suggested as benchmarks. There was a feeling among some Past Presidents that standards of scientific rigor in papers accepted for publishing and for presentation at conferences were not consistent. I believe that this is another issue that ISSS needs to look at in greater depth but not in isolation from the issue of relevance. The challenge for ISSS is to support ideas and individuals who bring both rigor and relevance into their work. Issues of relevance such as ethics and justice must be an equal measure of quality with scientific rigor.

5. Leadership – Systems Thinkers in Leadership Positions. One issue that stood out for me among many presentations was captured by the theme of Russell Ackoff's paper, "doing the right things right." This is the seminal challenge of the issue of leadership. For me the emergent issue is not one of how systems thinkers can influence leadership in organizations, governments and NGO's (Non-Governmental Organizations) through their compelling systems logic. For me the issue is how systems thinkers become leaders themselves. How do they transition from advising to serving? I believe that the issue of service is key to facilitating systems thinkers as leaders. Good leadership is not hierarchical but rather depends on the development of relationships among those serving and those being served. These relationships are made explicit in the form of contracts that are best formed through a systems approach. I would like to now discuss in greater depth the issue of **service** since it is the theme I have chosen as President of ISSS for the 45th annual conference.

Service

1. Congruence of behavior with systems ideas. The challenge for systems thinkers is to act congruently with espoused systems ideas. If one wishes to be influential as a change agent it is important to remember that actions speak louder than words or in the words of an American poet:

> "I can't hear what your saying because your actions speak so loud!"
> Ralph Waldo Emerson

Fundamental concepts like *open systems* behavior, lead to key questions of how ISSS can act so as to maintain open boundaries with the global socio-economic environment it is embedded in; how an open exchange is maintained with diverse groups of people who do not share the systems tradition of inquiry, but who can benefit from its insights. The idea of *unity* leads to the question of how reflection and action can be integrated through a systems approach rather than separated into different, distinct, and often competing domains of inquiry versus action. The systems concept of *relationship* also leads to fundamental questions about how systems scientists and practitioners relate to others who are not members of the systems science community, but are the focus of the systems change agent's intention. Because, description and explanation alone do not prescribe action and prediction and control do not alone justify action, it is necessary for systems thinkers and practitioners to create systemic relationships with those who desire change in their lives, but do not like being changed. This relationship, in the form of *service on behalf of the other*, is one that needs to be explored as a means for systems thinkers as leaders to act systemically and responsibly with full accountability for the consequences of their work whether intended or unintended.

2. Reintegration of Sophia. Philosophy is a compound word formed from two Greek terms, philo – love for- and sophia – wisdom. Thus philosophy is the love of wisdom. In the pre-Socratic era of Greek history sophia or wisdom was defined as the knowing hand. Thinking and acting or making were not divided into separate domains. But, during the Socratic period and as a consequence through all later periods of Western thought up to today, those who thought about things were put at the top of the hierarchy of important types of people while those who made things fell to the bottom of this hierarchy. This split in sophia is reflected in many of the polarities of modern life where we still live with the consequences. We have clear polarities between workers and managers, between idealists and realists, between researchers and practitioners— and the list goes on. From a systems perspective there is no logical reason for this split. Reflection and action need to be an integrated whole if systems ideas are to be congruent with systems behavior.

3. Systemic Relationships of Systems Thinkers to the Rest of the World. Systems thinkers live and work in a world populated by people who see and experience the world in very different ways most of whom are not from a systems science tradition. Yet from a systems perspective we are all part of the interconnected social systems we inhabit. Systems thinkers, in the spirit of congruence with systems ideas, need to form relationships of interconnectedness with these others who are often the focus of systems change agents who claim agency without community. The relationship I want to focus on is that of **service** from a systems perspective. Forming a service relationship is forming a social system with all the interconnected and emergent qualities of any complex system. It is a 'conspiracy' i.e. a breathing together of divergent individuals who by their differences define the rich potential of the social system.

4. Service as a Systemic Relationship

> "People like to change, they just don't like being changed"
> Chilean Student

Being in service does not mean being a servant or subservient. It does not mean acting as an expert or a mere facilitator on behalf of someone else's needs. Service is not about helping people create what they already know they want or imposing predetermined idealized solutions onto unique particular situations. The success of service can be best determined when those being served experience the surprise of self recognition in what emerges out of their interaction with systems thinkers, who have attended to broader systemic concerns while serving particular autonomous interests.

Service that is not servitude treats the others as an equal. This does not mean the same as being similar, as in categories of social science, or equivalent, as in egalitarianism, but equal in terms of the right to have anyone's desiderata become the seed for purposeful change. Service is distinct from helping, which by its nature creates a unilateral relationship. In this type of relationship all power and resources reside in a dominant role. Leaving the other in a position of being helpless and indebted:

> Serving is different from helping. Helping is based on inequality; it is not a relationship between equals.... Service is a relationship between equals.... Helping incurs debt. When you help someone they owe you one. But serving, like healing is mutual. There is no debt.
> Rachel Naomi Remen (1996)

Service is a relationship of mutuality and diversity allowing those with excellent skills and abilities to work on behalf of those who are pressing to express their humanity in ways which honor their uniqueness in a world of increasing globalization.

5. Conference Theme for 2001. The theme for the 45th annual conference therefore is this:

Systems Science in the Service of Humanity:

With the concomitant questions of:

1) How can Service Serve Development?
2) How can Systems Thinkers Serve?

Thank you for coming today and I very much look forward to being your President for the coming year.

INDEX

MIX
Papier aus verantwortungsvollen Quellen
Paper from responsible sources
FSC® C105338

If you have any concerns about our products,
you can contact us on
ProductSafety@springernature.com

In case Publisher is established outside the EU,
the EU authorized representative is:
Springer Nature Customer Service Center GmbH
Europaplatz 3, 69115 Heidelberg, Germany

Printed by Libri Plureos GmbH
in Hamburg, Germany